BIOCHEMISTRY RESEARCH TRENDS

METHANOGENESIS

BIOCHEMISTRY, ECOLOGICAL FUNCTIONS, NATURAL AND ENGINEERED ENVIRONMENTS

BIOCHEMISTRY RESEARCH TRENDS

Additional books in this series can be found on Nova's website under the Series tab.

Additional e-books in this series can be found on Nova's website under the e-book tab.

BIOCHEMISTRY RESEARCH TRENDS

METHANOGENESIS

BIOCHEMISTRY, ECOLOGICAL FUNCTIONS, NATURAL AND ENGINEERED ENVIRONMENTS

GAGIK BADALIANS GHOLIKANDI
EDITOR

nova publishers
New York

For permission to use material from this book please contact us:
Telephone 631-231-7269; Fax 631-231-8175
Web Site: http://www.novapublishers.com

Library of Congress Cataloging-in-Publication Data

ISBN: 978-1-63321-567-2

Library of Congress Control Number: 2014945323

Published by Nova Science Publishers, Inc. † New York

CONTENTS

PREFACE

The combustible gases have been emitted from natural and geological sources since ancient time. It was the experiments of Alessandro Volta on combustible air achieved from swampy and oozy areas that created interests to guide scientific researches according to these gases, and consequently, led to exploration and recognition of the methane gas as an important member of this group. The ancients, i.e., the Greeks, Persian, and Indian discovered natural gas. They were excited by the burning springs created when natural gas seeped from the ground cracks and was ignited by lightning. Sometimes they built temples around these everlasting flames and worshipped the fire. About 2,500 years ago, the Chinese recognized that they can use natural gas as a source of energy. The Chinese transported the gas by pipes from shallow wells and burned it under large pans to evaporate sea water for salt production.

Methane is one of the most effective gases that have affected human's life especially as a satisfactory source of energy. Nowadays, the industrial development and increasing demand to other energy sources than oil and coal brought methane gas into the front as a substitute for them. The most growing use of natural gas is for the electrical power generation. At atmospheric pressure one cubic meter of biogas is equivalent to about 0.6 liters of the fuel oil.

On the other hand methane, an important greenhouse gas, has built up rapidly in the atmosphere since the nineteenth century and is capable of harming or destroying the planet if there are uncontrolled emissions. A global increase in the occurrence of extreme weather events has led to an increasing awareness of the potential role of anthropogenic sources of greenhouse gases in climate change. The rapid urbanization and industrialization in recent decades, evidently predicts the requirement for efficient systems for conversion of the generated wastes such as agricultural, domestic and industrial solid waste, and wastewater into useful end products. The common preferred process for the waste sludge stabilization is the anaerobic digestion, due to the related energy recovery from sludge, i.e., biogas production and anaerobically-digested sludge application, i.e., agricultural and industrial uses. Consequently, the environmental problem caused by sewage sludge can be solved in a sustainable way; thus allows water and stabilized sludge to utilize for reuse and reclamation purposes. However, the anaerobic treatment process of wastewater and stabilization of waste sludge may also create an environmental problem, i.e., greenhouse gases uncontrolled emissions, especially carbon dioxide and methane. Bio-electrochemical systems are also capable to treat organic waste solids and wastewater, and generate electrical energy simultaneously. These technologies present many exciting opportunities and chronologically contain different challenges to be utilized in various applications.

These facts make us to take a more precise look at the need to recycle waste rather than simply find ways to dispose of it. Nowadays, the world is confronted with the twin challenges of fossil-fuel depletion and greenhouse gases emissions. Therefore, numerous scientists and engineers are obliged to investigate various solutions in order to optimize the methanogenesis process in different environments and to overcome the methane production and utilization problems.

However many attractive features can be offered by anaerobic systems, i.e., low operating and maintenance costs, biogas production, low energy and labor requirement, low sludge production, and high efficiency of organics removal; they unfortunately encountered unwarranted reputation as an unstable and difficult to control process. Owing to complexity of the anaerobic digestion process, there has been a growing demand to simulation and monitoring by means of different modeling approaches, i.e., mathematical modeling and software tools. Artificial neural network (ANN) is one of these methods that have been employed to simulate engineered environments performance in different operation conditions.

This book aims at the methanogenesis process occurrence from different points of view, describing its ecological functions and possible environmental impacts, providing an assessment of the existing related technologies and techniques, and highlighting various aspects of bio-energy generation in seven chapters.

In the first chapter a brief review to methanogenesis biochemistry and the sources of methane (natural and engineered environments) as well as its impacts on the environment has been presented. The second chapter discusses the nonbacterial methanogenesis mechanisms and its significance. The third chapter describes the methanogenesis process in a novel engineered environment, i.e., electrolysis-enhanced anaerobic baffled reactor. The modeling of anaerobic digestion by means of numerical methods and software tools has been proposed in fourth and fifth chapters. Chapter six discusses the anaerobic waste sludge digestion and the related affecting parameters. Finally, bio-electrochemical processes and techniques for bio-energy generation have been discussed in seventh chapter.

The main goal of this book is to gather different updated viewpoints according to the methanogenesis process from scientifically and technologically diverse areas in order to present students, researchers, managers and engineers with useful knowledge in this regard.

Editor:
Assoc. Prof. Dr. Gagik Badalians Gholikandi
Faculty of Water Engineering and Environment
Shahid Beheshti University, A. C., Tehran, Iran
g.badalians@yahoo.com

Co-editor:
Hamed Rasouli, MSc.
Faculty of Water Engineering and Environment
Shahid Beheshti University, A. C., Tehran, Iran

In: Methanogenesis
Editor: Gagik Badalians Gholikandi
ISBN: 978-1-63321-567-2

Chapter 1

AN OVERVIEW OF THE METHANOGENESIS PROCESS OCCURRENCE IN NATURAL AND ENGINEERED ENVIRONMENTS

Gagik Badalians Gholikandi [*]
and Hamed Rasouli Sadabad
Faculty of Water Engineering and Environment,
Shahid Beheshti University, A.C., Tehran, Iran

ABSTRACT

Anaerobic processes may be defined as a complicated system of multiple reactions resulting in the production of methane, i.e., methanogenesis process. Complex organic molecules such as polysaccharides, lipids, and proteins are degraded to simpler compounds through successive biochemical reactions by a variety of bacteria and archaea to methane. Microbial methane production occurs in a variety of conditions, resulted from natural processes or attributed to anthropogenic activities, i.e., engineered environments. Natural methane gas is utilized frequently as a clean energy source. It is rapidly gaining importance in global energy markets. On the other hand methane, an important greenhouse gas, has built up rapidly in the atmosphere since the nineteenth century. Numerous studies have been focused on investigating

[*] Corresponding author: Hakimieh- Tehranpars, Shahid Abbaspour Blvd., PO Code 16589-53571, PO Box 16765-1719, Tehran, Iran. Email: g.badalians@yahoo.com.

and determining the methanogenesis process mechanisms and characteristics, and the methane gas emissions according to various environmental sciences and engineering areas.

Keywords: methane, methanogenesis, emissions sources, energy source, greenhouse gas

1.1. INTRODUCTION

Biological processes consist of promoting and maintaining a microbial population, i.e., biomass that metabolizes a target matter (biodegradation). Thus, the biological treatment involves complex interaction of mixed biological populations. Methanogens or Methane-forming archaea are strictly anaerobic microorganisms, and are found in anoxic environments such as fresh water sediments, tundra areas, swamps and as well as anaerobic sludge digesters in wastewater treatment plants (Badalians Gholikandi 2006; Welte and Deppenmeier 2013). The methanogens form a phylogeneticcally diverse group, classified as methanobacteriales, methanococcales, methanomicrobiales, methanosarcinales and methanopyrareles, divided into 10 families and 31 genera (Garcia et al., 2000; Liu and Whitman 2008; Welte and Deppenmeier 2013).

Anaerobic processes may be defined as a complicated system of multiple reactions resulting in the production of methane, i.e., methanogenesis process or biomethanation. Decomposition of organic matter to methane and carbon dioxide in an anaerobic environment requires the syntrophic collaboration of anaerobic bacteria and methanogenic archaea (Liu 2010; Whitman 2008).

Methanogenic substrates can be divided into three major groups as follows (Liu and Whitman 2008; Thauer et al., 2008):

H_2/CO_2, formate, and carbon monoxide: The hydrogenotrophic methanogens can reduce the CO_2 to CH_4 using H_2 as electron donor. A large number of the methanogens can also use formate or carbon monoxide in this regard. By the process of hydrogenotrophic methanogenesis, the conversion of CO_2 into CH_4 occurs consecutively on formyl, methylene and methyl levels. Special coenzymes such as Methanofuran (MFR), Tetra-hydromethanopterin (H_4MPT) and coenzyme M (HS-CoM) are necessary for the occurrence of the related coupling reactions (Liu and Whitman 2008; Thauer et al., 2008; Stamps and Plugge 2010).

Methanol and methyl compounds: Methanosarcinales and methanosphaeras convert these substrates to methane gas. The methyl group of these compounds is first transferred to a corrinoid protein; thereafter it moves to a HS-CoM enzyme. The Methyl-CoM enters the methanogenesis pathway afterwards and consequently the reduction reaction to the methane gas occurs. The required electrons are obtainable through the oxidation of methyl-CoM to CO_2. This may be defined as a reverse reaction of the hydrogenotrophic methanogenesis process (Liu and Whitman 2008; Sprenger et al., 2005; Lange and Ahring 2001; Welte and Deppenmeier 2013).

Acetate: Acetate is the main intermediate of the anaerobic food chain. Acetate, i.e., acetic acid can undergo a dismutation reaction to produce methane and carbon dioxide. Methanosarcina and methanosaeta consume acetate as the methanogenic substrate. Methanosarcina utilizes the low-affinity acetate kinase/phosphor-transacetylase to form acetyl-CoA while methanosaeta uses the high-affinity acetyl-CoA synthase which requires a higher energy savings than acetate kinase. It splits into methyl-CoM and carbon monoxide; the reduction of methyl-CoM to CH_4 and the oxidation of CO to CO_2 occur afterward (Jetten et al., 1992; Welte and Deppenmeier 2013).

Methane gas is a colorless, odorless, and flammable hydrocarbon. It is a tetrahedral molecule with four equivalent C-H bonds. The formation of carbon dioxide, water, and energy occurs as the result of the methane gas combustion in the presence of a sufficient concentration of oxygen. Because of the high energy content, methane gas is becoming the fuel of choice for a wide array of uses (Chinoweth 1996). On the other hand, the global warming potential (GWP) of the methane gas molecule over a period of 100 years is 25 times greater than a molecule of carbon dioxide (Dlugokencky et al., 2009; Reay et al., 2010).

The chemical lifetime of the methane is dependent on its removal reactions with the hydroxyl radicals (OH°) and is estimated to be about 9.6 years (Folland et al., 2001). The perturbation lifetime of the methane molecule equals about 12 years. The rise in the methane emissions reduces the concentration of hydroxyl radicals and consequently may increase the methane lifetime in the atmosphere (Solomon et al., 2007a; Amstel et al., 2010). Variations in the atmospheric methane concentration and hence water vapor in the stratosphere leading to the climate change and the change of optical thickness of polar clouds in the stratosphere (Blake and Rowland 1988; Thomas et al., 1989; Sloan et al., 1992; Sloan and Pollard 1994; Lelieveld et al., 1998; Hansen and Sato 2004; Hansen et al., 2007; Beerling et al., 2009). Ice core records indicate that over the last 650,000 years, the atmospheric methane concentration has been increased from lower than about 400 ppb during glacial periods to more than about 700 ppb during interglacial periods. Since the beginning of the industrial era the concentration of atmospheric methane has been increased to about 1,000 ppb in the late 1700s. In 2005, the global average methane concentration was measured in both hemispheres equals 1,774.62 ± 1.22 ppb (Forster et al., 2007; Dlugokencky et al., 2009; Amstel et al., 2010).

1.2. Methane Formation in Natural and Engineered Environments

Nowadays, while scientific information shows various potentially devastating problems with global ecosystems, methanogenesis process attract considerable attention. The problem with methane emissions sources has a long history. On the other hand, the methane fermentation has been used for different industrial and environmental protection purposes.

According to the reports published in 2008 by the Inter-governmental Panel on Climate Change (IPCC), the total flux rate of the methane gas emissions into the atmosphere equals currently about 566 teragrams per year ($TgCH_4$/year)(Solomon et al., 2007b, Stamps and Plugge 2010). The natural sources are currently the cause of 208 $TgCH_4$/year or 37 percent of total global emissions into the atmosphere. The remainder can be determined as a result of anthropogenic sources emissions (Forster et al., 2007; USEPA 2010; Stamps and Plugge 2010).

1.2.1. Natural Environments

Methane gas natural sources emit noticeable amounts of the methane into the atmosphere every year. Wetlands, vegetation, biomass burning, ruminants, oceans, estuaries, rivers, permafrost, lakes, gas hydrates and geological sources are the important related natural sources.

Wetlands: Various parameters like temperature, water level, available organic matter and their characteristics affect the biological activity of different types of methane-forming bacteria in wetlands soil body; noticeable changes in flux rate occurs by a relatively limited variation of environmental parameters. There are three major pathways through which the methane emissions occur, i.e., bubbling, moving along the soil layers, and moving through the plants. According to the related reports, the flux rates are in the range from 24 to 72 Tg CH_4/yr for high latitude wetlands, with an average of 43 Tg CH_4/yr. Reported emissions quantity pertaining to tropical wetlands varies from 81 to 206 Tg CH_4/yr, with an average of 128 Tg CH_4/yr (Christensen 2010). The impacts of climate change in the 21st century on these emissions could be significant (Shindell et al., 2004).

Vegetation: Keppler et al. (2006) estimated that the approximate quantity of the related methane emissions might be 10 to 40 percent of the total global emissions. Another studies (Lowe 2006; Schiermeier 2006) proposed that the benefits of forests might be substantially reduced to store carbon deoxide and carbon. Even at an early stage, two reports (Bergamaschi et al., 2006; Nieps 2006) proposed that the methane produced by plants is only a small fraction of the global methane budget and seems to be insignificant comparing with benefits from carbon sequestration. Recent studies showed that the aerobic methane formation is an integral part of cellular responses towards changes in oxidative status in all eukaryotes. It is probable that some parts of the emitted methane by plants originate from soil processes via the transpiration stream or physical diffusion through plant tissues. It seems that the plants generate methane due to environmental stresses. In other words, the other feasible emissions are originated from tempting biomolecules by ambient UV light, elevated temperatures, hypoxia, or from more general physiological processes. More researches are needed to determine the contribution of stress-induced emissions to the global methane budget, considering the methane consumption by the methanotrophs (Christensen 2010).

Biomass burning: Biomass burning is a significant emissions source of gases into the regional and global atmosphere (Crutzen et al., 1979; Seiler and Crutzen 1980; Crutzen and Andreae 1990; Levine et al., 1995; Christensen 2010). The bulk of the world's biomass burning occurs in the tropics such as the tropical forests of South America and Southeast Asia as well as the savannas of Africa and South America. It is estimated that biomass burning releases 14 to 88 Tg CH_4/year into the atmosphere (Christensen 2010).

Ruminants: Ruminant livestock usually produces methane as a by-product of the feed fermentation in their rumens. The bulk, i.e., more than 90 percent of the methane is emitted to the atmosphere through belching. Some dairy cattle emit several hundreds of liters of the methane daily. In 2005, the methane emissions from ruminant livestock was estimated to be around 75 Tg-CH_4/year (Kelliher and Clark 2010).

Oceans, Estuaries, and Rivers: In water bodies, the methane gas is usually produced by microbial processes both in the water column and in sediments. The results of the measurements conducted in 1960s and 1970s showed that generally, surface waters are relatively small sources of the methane emissions. Methane production is greater in areas under freshwater and in shallow water containing organic sediments. An average of recent estimates gives a total flux of 5.5 $TgCH_4$/year. Estuaries and rivers cover limited areas comparing with seas or oceans, yet they contain much biological activity. They are also sites of active mixing and enabling methane production in nearby wetlands and shallow-water environments. The produced methane releases rapidly into the atmosphere. Rivers and estuaries approximately release 1.85 Tg CH_4/year (US EPA 2010).

Permafrost: Permafrost is widespread in the arctic and also exists intermittently in the sub-arctic and boreal regions. US EPA's report (1993) included the permafrost as a natural source of the methane because early results showed that the frozen methane gas can be released when this layer melts. The methane gas emissions are estimated to be 0 to 1 $TgCH_4$/year (USEPA 2010). The methane gas concentration in the permafrost is not a considerable number. it must pass through the overlying defrosted soil before releasing into the atmosphere. During this transfer process, a portion or most of the methane will be oxidized to carbon dioxide. The thermokarst erosion can form or dry lakes or wetlands up, thus affects the methane gas emissions rate from high latitude lakes and wetlands.

Lakes: The methane gas production rate in anoxic sediments depends on temperature, available organic matter and oxygen-free condition. There are four pathways for methane gas emission, i.e., Bubbling, Diffusion, Plant-mediated transport, and seasonal overturning. Flux rates of the pathways generally rise according to increasing wind speed. Regarding recent estimates, the lakes emissions equals about 30 $TgCH_4$/year (US EPA 2010).

Gas Hydrates: The gas hydrates crystals formation occurs between water and gaseous phase under high pressure and in ambient temperature. A considerable quantity of the hydrated methane gas is in the continental shelf and permafrost region. The huge quantity of methane gas in the continental shelf is so overwhelming that only a few percentage of the methane released could lead to a chain reaction of the planet's overheating. The wide spewing of the hydrated methane gas can be limited for various reasons, i.e., oxidation or adsorption in the sediment and finally dissolve in the water body. Gas hydrates are sensitive to climate change. The Effect of the temperature gradient is more significant than the pressure; the methane gas emissions are believed to be increased as temperature rises (US EPA 2010).

Geological Sources of Methane: The methane gas is one of the several gases produced by geological processes as well as CO_2, H_2, H_2S, N_2, SO_2, H_2O (as steam or water vapor), and the petroleum gases (ethane, propane, butane and pentane). The methane gas formation occurs in various geological environments which are categorized as biogenic, thermogenic, and abiogenic (Judd 2000). The estimation of methane gas bottom-up emission generally ranges from 32 to 74 $TgCH_4$/year for both marine and terrestrial sources (Etiope 2010).

1.2.2. Methane in Engineered (Anthropogenic) Environments

Rice cultivation, solid waste stabilization and wastewater treatment processes, and operations which are related to the fossil fuels extraction or transportation are major anthropogenic sources of the methane emissions.

Rice cultivation: Rice paddies as an important methane gas emissions source have been investigated comprehensively, e.g., Wassmann et al. (1993, 1997, 1998), Houghton et al. (1995) and Olivier et al. (1996). The frequently saturated carbon-rich soil of the rice paddy in anoxic condition is a suitable environment for a high rate occurrence of the microbial methanogenesis. Considering the recent scientific information, rice cultivation causes methane gas emissions equal 25 to 50 $TgCH_4$/year (Conen et al., 2010; Fazli et al., 2013).

Solid waste and wastewater treatment: Biological treatment process is the degradation of organic matter by the action of microorganisms. Anaerobic processes have been used for the municipal and industrial wastewaters treatment and waste sludge stabilization for over a century. These processes convert organic matter into methane and carbon dioxide, called biogas (McCarty and Smith 1986). The approximate quantity of the methane emissions from agricultural waste and wastewater equals 14 and 25 Tg-CH_4/year (Miriam et al., 2010). The production of biogas is an advantage over biological processes due to its feasible reuse as an energy source. The energy content of the biogas depends on its methane content. Biogas is used as fuel for combustion engines, which convert it into mechanical energy, powering an electric generator to produce electricity. Biogas plants are able to convert animal manure, green plants, agro industries waste and slaughterhouses into combustible gas. However, attention should be taken to avoid the undesired emissions of greenhouse gases such as CH_4 and N_2O into the atmosphere (US EPA 2010). This bacterial process consists of four successive reaction steps, i.e., hydrolysis, acidogenesis, acetogenesis, and methanogenesis.

Anaerobic digestion can be performed in a single-stage or two-stage digester. In the two-stage digestion process, two digesters in series separate the functions of biological stabilization from gravity thickening and storage (Hammer and Hammer 2008). There are three groups of anaerobic bacteria with respect to the temperature that they remain active, i.e., psychrophiles, mesophiles, and thermophiles (Gerardi 2006). The methane gas emissions from leaky digesters can be considerable.

The sewage collection and transport system is one of the main sources of the methane gas emissions. Open sewers, networks of the open canals, gutters and ditches are considerable methane emissions sources (IPCC 2006d; Miriam et al., 2010).

Landfill: Landfill sites provide ideal conditions for methanogens because of available biodegradable substrates which are held under anoxic conditions. These unique properties made landfill a potent source of the methane gas production and related emissions. The landfill produced methane gas is supposed to approximately be 1.3 percent of the global anthropogenic greenhouse gas emissions (Monni et al., US EPA 2010; Bogner et al., 2007; Rogner et al., 2007; Bogner and Spokas 2010). The transfer and emitting mechanisms of the formed methane from landfills into the atmosphere are diffusion, advection, ebullition (bubbling) and plant-mediated transport. Similar to the wetland ecosystem, diffusion is the major mechanism of the methane gas emissions (Bogner et al., 1987; Chanton 2005). The collected gas from landfill contains up to 60 percent methane gas. Therefore, it is an important local source of renewable energy for various applications. Optimizing the activity of existing methanotrophic microorganisms in the landfill's soil cover causes a decrease in the methane gas emissions quantity (Scheutz et al., 2003; Barlaz et al., 2004; Bogner et al., 2010).

Manure: For various reasons, i.e., available biodegradable substances, existing moisture, and impressive ecological condition, manure is an appropriate environment for the methanogens microbial activity. The biogas emissions are the result of anaerobic fermentation process. Methane production is often increased by means of co-digestion of the manure with other biomass (Bogner and Spokas 2010).

Fossil fuel sources: The methane gas emissions from fossil fuel sources include natural gas losses, oil related emissions and coal bed methane. A considerable part of the methane emissions in this regard derives from methane gas release during the fossil fuel extraction, storage and handling. A few quantities of the methane emissions occur during the incomplete fossil fuel combustion. The estimated global methane gas emissions rate from fossil fuel sources is currently between 74 and 77 $TgCH_4$/year. This is excluded the emissions from biomass burning (Mattus and Kallatrand 2010). Large quantities of methane release while coal mining and natural gas extracting into the atmosphere. While drilling for oil and during the subsequent extraction processes, the trapped methane gas release into the atmosphere in large quantities. The major part of the methane which has been generated during the coal formation process remains trapped within the coal seam and the surrounding strata. Mining operations cause the release of the trapped gas into the atmosphere (US EPA 2006).

REFERENCES

van Amstel, A., Reay, D. (2010). Methane and climate change, edited by Reay, D., Smith, P. & van Amstel, A., earthscan publishing, London, Washington, DC, 240-246.

Andreae, M. O. (1991). Biomass burning: Its history, use, and distribution and its impact on environmental quality and global climate, in J. S. Levine (ed) Global Biomass Burning: Atmospheric, Climatic, and Biospheric Implications, MIT Press, Cambridge, MA: 3-21.

Andreesen, J. R. (1994). Glycine metabolism in anaerobes. *Antonie van Leeuwenhoek, 66,* 223-237.

Badalians Gholikandi, G. (2006). Applied Microbiology of Water and Wastewater. Nopardazan Publ, Tehran, Iran.

Badalians Gholikandi, G. (2009). Wastewater treatment plant design. Ayeej publication, Tehran, Iran.

Badalians Gholikandi, G., Moradhasseli, M. & Riahi, R. (2009). Treatment of domestic wastewater in a pilot-scale HSFCW in West Iran. *Desalination, 248,* 977-987.

Barker, H. A. (1981). Ammino Acid Degradation by Anaerobic Bacteria. *Annual Review of Biochemistry, 50,* 23-40.

Barlaz, M. A., Green, R. B., Chanton, J. P., Goldsmith, C. D. & Hater, G. R. (2004). Evaluation of a biologically active cover for mitigation of landfill gas emissions. *Environmental Science and Technology, 38,* 4891-4899.

Beerling, D. J., Brentnall, S. J. (2007). Numerical evaluation of mechanisms driving Early Jurassic changes in global carbon cycling. *Geology, 35,* 247–250.

Beerling, D. J., Gardiner, T., Leggett, G., Mcleod, A. & Quick, W. P. (2008). Missing methane emissions from leaves of terrestrial plants. *Global Change Biology, 14(8),* 1821-1826.

Bergamaschi, P., Dentener, F., Grassi, G., Leip, A., Somogyi, Z., Federici, S., Seufert, G. & Raes, F. (2006). Methane emissions from terrestrial plants. European Commission, DG Joint Research Centre, Institute for Environment and Sustainability, Ispra, Italy.

Blackburn, J. W., Cheng, J. (2005). Heat production profiles from batch aerobic thermophilic processing of high strength swine waste. *Environmental Progress, 24,* 323-333.

Blaut, M. (1994). Metabolism of methanogens. *Antonie Van Leeuwenhoek, 66(1-3),* 187-208.

Bogner, J., Abdelrafie-Ahmed, M., Diaz, C., Faaij, A., Gao, Q., Hashimoto, S., Mareckova, K., Pipatti, R. & Zhang, T. (2007). Waste Management. In Metz, B., Davidson, O. R., Bosch, P. R., Dave, R. & Meyer, L. A. (eds) Climate Change 2007, mitigation contribution of working group III to the fourth assessment report of intergovernmental panel on climate change, Cambridge University press, Cambridge, UK and New York, NY.

Bogner, J., Chanton, J., Blake, D., Abichou, T. & Powelson, D. (2010). Effectiveness of a Florida landfill biocover for reduction of CH4 and NMHC Emissions. *Environmental Science and Technology.*

Bogner, J., Lagerkvist, A. (1997). Organic carbon cycling in landfills: models for a continuum approach. *In Proceedings of Sardinia 97 International Landfill Symposium, published by CISA,* University of Cagliari, Sardinia.

Bogner, J., Spokas, K. (2010). Methane and climate change, edited by Reay, D., Smith, P. & van Amstel, A., earthscan publishing, London, Washington, DC, 175-200.

Bowen, G. J., Bralower, T. J., Delaney, M. L., Dickens, G. R., Kelly, D. C., Koch, P. L., Meng, J., Sloan, L. C., Thomas, E. C., Wing, S. L. & Zachos, J. C. (2006). Eocene hyperthermal event offers insights into greenhouse warming: American Geophysical Union, *EOS, Transactions, 87,* 165–169.

Chanton, J. P. (2005). The effect of gas transport on the isotope signature of methane in wetlands. *Organic Chemistry, 36,* 753-768.

Chappellaz, J. A, Fung, I. Y. & Thompson, A. M. (1993). The atmospheric CH4 increase since the last glacial maximum: 1. Source estimates. *Tellus, 45B,* 228-241.

Chinoweth, D. P. (1996). Environmental impacts of biomethanogenesis. *Environmental Monitoring and Assessment, 42(1-2),* 3-18.

Christensen, T. R. (2010). Methane and climate change, edited by Reay, D., Smith, P. & van Amstel, A., earthscan publishing, London, Washington, DC, 27-41.

Conen, F., Smith, K. A. & Yagi, K. (2010). Methane and climate change, edited by Reay, D., Smith, P. & van Amstel, A., earthscan publishing, London, Washington, DC, 114-135.

Crutzen, P. J., Andreae, M. O. (1990). Biomass burning in the tropics: Impact on atmospheric chemistry and biogeochemical cycles. *Science, 250,* 1669-1678.

Crutzen, P. J., Sanhueza, E. & Brenninkmeijer, C. A. M. (2006). Methanerroduction from mixed tropical savanna and forest vegetation in Venezuela. *Atmospheric Chemistry and Physics Discussions, 6,* 3093-3097.

Daniels, L., Sparling, R. & Sprott, G. D. (1984). The bioenergetics of methanogenesis. *Biochemistry Biophys. Acta, 768(2),* 113-163.

De Bok, F. A. M., Stams, A. J. M., Dijkema, C. & Boone, D. R. (2001). Pathway of propionate oxidation by a syntrophic culture of Smithella propionica and Methanospirillum hungatei. *Applied and Environmental Micobiology, 67,* 1800-1804.

De Mes, T. D. Z., Stams, A. J. M., Reith, J. H. & Zeeman, G. (2003). Methane production by anaerobic digestion of wastewater and solid wastes, in J. H. Reith, R. H. Wijffels and H. Barten (eds) *Bio-methane and Bio-hydrogen: Status and Perspectives of Biological Methane and Hydrogen Production, Dutch Biological Hydrogen Foundation,* Energy Research Centre of The Netherlands, 58-102.

Deppenmeier, U. (2002). The unique biochemistry of methanogenesis. Prog Nucleic Acid. *Research Mol. Biology, 71,* 223-283.

Dickens, G. R., O'Neil, J. R., Rea, D. K. & Owen, R. M. (1995). Dissociation of oceanic methane hydrate as a cause of the carbon isotope excursion at the end of the Paleocene. *Paleoceanography, 10,* 965–971.

Dlugokencky, E. J., Bruhwiler, L., White, J. W. C., Emmons, L. K., Novelli, P. C., Montzka, S. A., Masarie, K. A., Lang, P. M., Crotwell, A. M., Miller, J. B. & Gatti, L. V. (2009). Observational constraints on recent increases in the atmospheric CH4 burden. *Geophysical Research Letters, 36,* L18803.

Etiope, G. (2010). Methane and climate change, edited by Reay, D., Smith, P. & van Amstel, A., earthscan publishing, London, Washington, DC, 42-62.

Fazli, P., Man, H. C., MdShah, U. & Idris, A. (2013). Characteristic of methanogens and methanotrophs in rice fields. *Journal of Biology and Biotechnology*, 21, 3-17.

Fonknechten, N., Chaussonnerrie, S. & Tricot, S. (2010). Clostridium sticklandii, a specialist in amino acid degradation: revisiting its metabolism through its genome sequence. *BMC Genomics,*France.

Forster, P., Ramaswamy, V., Artaxo, P., Berntsen, T., Betts, R., Fahey, D. W., Haywood, J., Lean, J., Lowe, D. C., Myhre, G., Nganga, J., Prinn, R., Raga, G. M. S. & Van Dorland, R. (2007). Changes in atmospheric constituents and in radioactive forcing, in S. Solomon, D. Qin, M. Manning, Z. Chen, M. Marquis, K. B. Averyt, M. Tignor and H. L. Miller (eds) Climate Change 2007: The physical science basis. Contribution of working Groupe to the fourth assessment report of Intergovermental Panel on Climate Change, Cambridge University press, Cambridge, UK and NY.

Garcia, J. L., Patel, B. K. C. & Ollivier, B. (2000). Taxonomic, Phylogenetic, and Ecological Diversity of Methanogenic Archaea, *Anaerobe 6,* 205-226.

Gedney, N., Cox, P. M. & Huntingford, C. (2004). Climate feedback from wetland methane emissions. *Geophysical Research Letters, 31,* L20503.

Gerardi, M.H. (2006). Wastewater bacteria. Wastewater Microbiology Series, Wiley & Sons, Inc., Hoboken, New Jersey.

Gottschalk, G. (1985). Bacterial Metabolism. 2nd Edition, Springer Verlag, New York.

Hammer, M. J., Hammer, M. J. Jr. (2008). Water and Wastewater Technology. Sixth Edition, Pearson Prentice Hall™Publishing, NJ Columbus, Ohio, 431-435.

Hansen, J., Sato, M. (2004). Greenhouse gas growth rates. *Proceedings of the National Academy of Sciences, USA, 101(16),* 109 –114.

Hansen, J., Sato, M., Kharecha, P., Russell, G., Lea, D. W. & Siddall, M. (2007). Climate change and tracegases. *Philosophical Transactions of the Royal Society A, 365,*1925–1954.

Ianotti, E. L., Kafkewitz, D., Wolin, M. J. & Bryant, M. P. (1973). Glucose fermentation products by Ruminococcus albus grown in continuous culture with Vibrio succinogenes: Changes caused by interspecies transfer of H2. *Journal of Bacteriology, 114,* 1231-1240.

IPCC (Intergovernmental Panel on Climate Change). (2006). Guidelines for National Greenhouse Gas Inventories, S. Eggleston, L. Buendia, K. Miwa, T. Ngara and K. Tanabe (eds), Institute for Global Environmental

Strategies, Kanagawa, Japan, www.ipcc-nggip.iges.or.jp and www.ipcc.ch.

Jetten, M. S. M., Stams, A. J. M. & Zehnder, A. J. B. (1992). Methanogenesis from acetate: A comparison of the acetate metabolism in Methanothrix soehngenii and Methanosarcina spp. *FEMS Microbiological Reviews, 88,* 181-198.

Judd, A. G. (2000). Geological Sources of Methane. Atmospheric Methane, Springer, 280-303.

Judd, M. J., Kelliher, F. M., Ulyatt, M. J., Lassey, K. R., Tate, K. R., Shelton, I. D., Harvey, M. J. & Walker, C. F. (1999). Net methane emissions from grazing sheep. *Global Change Biology, 5,* 647-657.

Keppler, F., Hamilton, J. T. G., Brass, M. & Rockmann, T. (2006). Methane emissions from terrestrial plants under aerobic conditions. *Nature, 439(7073),* 187-191.

Keppler, F., Rockmann, T. (2007). Methane, plants and climate change. *Scientific American, 296(2),* 52-57.

Laanbroek, H. J. (2010). Methane emission from natural wetlands: interplay between emergent macrophyts and soil microbial processes. A mini-review. *Annals of Botany 105(1),* 141-153.

Laanbroek, H. J., Abee, T. & Voogd, I. L. (1982). Alcohol conversions by desulfobulbus propionicus Lindhorst in the presence and absence of sulphate and hydrogen. *Archives of Microbiology, 133,* 178-184.

Lange, M., Ahring, B. K. (2001). A comprehensive study into the molecular methodology and molecular biology of methanogenic Archaea, *FEMS Microbiology Reviews 25,* 553-571.

Lehr, J. H., Keeley, J. & Lehr J. (2005). Water Science Encyclopedia: Domestic, Municipal, and Industrial Water Supply and Waste Disposal, J. Wiley and Sons, New York.

Lelieveld, J., Crutzen, P. J. & Dentener, F. J. (1998). Changing concentration, lifetime and climate forcing of atmospheric methane. *Tellus, 50B,* 128-150.

Lettinga, G. (1995). Anaerobic digestion and wastewater treatment systems. *Antonie van Leeuwenhoek, 67,* 3-28.

Levine, J. S., Cofer, W. R., Cahoon, D. R. & Winstead, E. L. (1995). Biomass burning: A driver for global change. *Environmental Science and Technology, 29,* 120A-125A.

Liu, Y. (2010). Taxonomy of methanogens; handbook of hydrocarbon and lipid microbiology, Springer, 547-558.

Liu, Y., Whitman, W. B. (2008). Metabolic, phylogenetic, and ecological diversity of methanogenic archaea. *Annals of the New York Academy of Science,* 171-189.

Lowe, D. C. (2006). Global change: A green source of surprise. *Nature, 439(7073),* 148-149.

Mc Carty, P. L., Smith, D. P. (1986). Anaerobic wastewater treatment. *Environmental Science & Technology, 20(12),* 1200-1206.

Mattus, R., Kallstrand A. (2010). Methane and climate change, edited by Reay, D., Smith, P. & van Amstel, A., earthscan publishing, London, Washington, DC, 201-210.

McInerney, M. J. (1988). Anaerobic hydrolysis and fermentation of fats and proteins. In A. J. B. Zehnder (ed), Biology of anaerobic microorganisms, John Wiley & Sons, New York, 373-415.

McInerney, M. J., Bryant, M. P., Hespell, R. B. & Costerton, J. W. (1981). Syntrophomonas wolfei gen. nov. sp. nov, an anaerobic syntrophic, fatty acidoxidizing bacterium. *Applied and Environmental Microbiology, 41,* 1029-1039.

McInerney, M. J., Struchtemeyer, C. G., Sieber, J., Mouttaki, H., Stams, A. J. M., Schink, B., Rohlin. L. & Gunsalus, R. P. (2008). Physiology, ecology, phylogeny, and genomics of microorganisms capable of syntrophic metabolism. *Annals of the New York Academy of Sciences, 1125,* 58-72.

Miriam, E., Dooren, H., Lexmond, M. & Zeeman, G. (2010). Methane and climate change, edited by Reay, P., Smith, P. & van Amstel, A., earthscan publishing, London, Washington, DC, 151-174.

Monni, S., Pipatti, R., Lehtila, A., Savolainen, I & Syri, S. (2006). Global Climate Change Mitigation Scenarios for Solid Waste Management. VTT Publications, Technical Research Centre of Finland, Espoo.

Muller, N., Griffin, B. M., Stingl, U. & Schink, B. (2008). Dominant sugar utilizers in sediment of Lake Constance depend on syntrophic cooperation with methanogenic partner organisms. *Environmental Microbiology, 10,* 1501-1511.

Nagase, M., Matsuo, T. (1982). Interaction between amino-acid degrading bacteria and methanogenic bacteria in anaerobic digestion. *Biotechnology and Bioengineering, 24,* 2227-2239.

NIEPS (Nicholas Institute for Environmental Policy Solutions). (2006). Do recent scientific findings undermine the climate benefits of carbon sequestration in Forests? An expert review of recent studies on methane emissions and water trade offs. Duke University, Nicholas Institute for Environmental Policy Solutions, Durham, NC.

Oerlygsson, J., Houwen, F. P. & Svensson, B. H. (1995). Thermophilic anaerobic amino acid degradation: deamination rates and end-product formation. *Applied Microbiology and Biotechnology, 43(2),* 235-241.

Park, J., Park, S. & Kim, M. (2014). Anaerobic degradation of amino acids generated from the hydrolysis of sewage sludge. *Environmental Technology, 35(9),*1133-1139.

Reay, D., Smith, P. & Amstel, A. V. (2010). Methane and climate change, edited by Reay, D., Smith, P. & van Amstel, A., earthscan publishing, London, Washington, DC, 1-13.

Ramsay, I. R., Pullammanappallil, P. C. (2001). Protein degradation during anaerobic wastewater treatment: derivation of stoichiometry. *Biodegradation, 12(4),* 247-256.

Reilly, J., Prinn, R., Harisch, J., Fitzmaurice, J., Jacoby, H., Kicklighter, D., Melillo, J., Stone, P., Sokolov, A. & Wang, C. (1999). Multi-gas assessment of the Kyoto Protocol, *Nature, 401,* 549-555.

Rogner, H. H., Zhou, D., Bradley, R., Crabbé, P., Edenhofer, O., Hare, B., Kuijpers, L. & Yamaguchi, M. (2007). Introduction. In Metz, B., Davidson, O. R., Bosch, P. R., Dave, R. & Meyer, L. A. (eds). Climate Change 2007: Mitigation. Contribution of Working Group III to the Fourth Assessment Report of the Intergovernmental Panel on Climate Change, Cambridge University Press, Cambridge, UK, New York, 95-116.

Scheutz, C., Bogner, J., Chanton, J., Blake, D., Morcet, M. & Kjeldsen, P. (2003). Comparative oxidation and net emissions of methane and selected non-methane organic compounds in landfill cover soils. *Environmental Science and Technology, 37,* 5150-515.

Schiermeier, Q. (2006). The methane mystery. *Nature, 442,* 730-731.

Schink, B., Stams, A. J. M. (2006). Syntrophism among prokaryotes. In Workin, M. D., Falkow, S., Rosenberg, E., Schleifer, K. H. & Stackebrandt, E. (eds). The Prokaryotes: An Evolving Electronic Resource for the Microbiological community, 3rd Edition, Springer-Velrag, New York, 309-335.

Seiler, W., Crutzen, P. J. (1980). Estimates of gross and net fluxes of carbon between the biosphere and the atmosphere from biomass burning. *Climatic Change, 2*, 207-247.

Shindell, D. T., Walter, B. P. & Faluvegi, G. (2004). Impacts of climate change on methane emissions from wetlands. *Geophysical Research Letters, 31,* L21202.

Sloan, L. C., Pollard, D. (1994). Polar stratospheric clouds: a high latitude warming mechanism in an ancient greenhouse world. *Geophysical Research Letters, 25,* 3517–3520.

Sloan, L. C., Walker, J. C. G., Moore, T. C., Rea, D. K. & Zachos, J. C. (1992). Possible methane-induced polar warming in the early Eocene. *Nature, 357,* 320–322.

Solomon, D., Qin, M., Manning, Z., Chen, M., Marquis, K. B., Averyt, M., Tignor & Miller, H. L. (eds) climate change 2007, The Physical Science Basis. Contribution of Working Group I to the Fourth Assessment Report of the Intergovernmental Panel on Climate Change, Cambridge University Press, Cambridge and New York, 499-587.

Sousa, D. Z., Smidt, H., Alves, M. M. & Stams, A. J. M. (2009). Ecophysiology of syntrophic communities that degrade saturated and unsaturated long-chain fatty acids. *FEMS Microbiology Ecology, 68,* 257-272.

Spokas, K., Bogner, J. (2010). Limits and dynamics of methane oxidation in landfill cover soils. *Journal of Waste Management.*

Sprenger, W. W., Hackstein, J. H. & Keltjens, J. T. (2005). The energy metabolism of Methanomicrococcus blatticola, Physiological and biochemical aspects. *Antonie van Leeuwenhoek, 87,* 289-299.

Stams, A. J. M. (1994). Metabolic interactions between anaerobic bacteria in methanogenic environments. *Antonie van Leeuwenhoek, 66,* 271-294.

Stams, A. J. M., Plugge, C. M. (2010). Methane and climate change, edited by D. Reay, P. Smith and A. van Amstel, earthscan publishing, London, Washington, DC, 14-26.

Stams, A. J. M., Plugge, C. M. (2009). Electron transfer in syntrophic communities of anaerobic bacteria and archaea. *Nature Review Microbiology, 7,* 568-577.

Svensen, H., Planke, S., Malthe-Sorenssen, A., Jamtveit, B., Myklebust, R., Eidem, T. R. & Rey, S. S. (2004). Release of methane from a volcanic basin as a mechanism for initial Eocene global warming. *Nature, 429,* 542–545.

Thauer, R. K., Kaster, A. K., Seedorf, H., Buckel, W. & Hedderich, R. (2008). Methanogenic archaea: Ecologically relevant differences in energy conservation. *Nature Reviews Microbiology, 6,* 579-591.

USEPA (United States Environmental Protection Agency). (2010). Methane and Nitrous Oxide Emissions from Natural Sources, Report no. EPA-430-R-09-025, US Environmental Protection Agency, Office of Atmospheric Programs, Climate Change Division, Washington, DC.

Van Kuijk, B. L. M., Schlosser, E. & Stams, A. J. M. (1998). Investigation of the fumarate metabolism of the syntrophic propionate-oxidizing bacterium strain MPOB. *Archives of Microbiology, 169,* 346-352.

Welte, C., Deppenmeier, U. (2013). Review: Bioenergetics and anaerobic respiratory chains of aceticlastic methanogens, *Journal of Biochimica et Biophysica Acta*, (in press).

In: Methanogenesis
Editor: Gagik Badalians Gholikandi

ISBN: 978-1-63321-567-2

Chapter 2

NONBACTERIAL BIOTIC METHANOGENESIS, POSSIBLE MECHANISMS AND SIGNIFICANCE

Eszter Tuboly, András Mészáros and Mihály Boros*
Institute of Surgical Research, University of Szeged, Hungary

ABSTRACT

Methane was earlier commonly thought to be produced in the gastrointestinal tract exclusively by methanogenic bacterial fermentation, under strictly anaerobic conditions. This notion was challenged when various in vitro and in vivo studies revealed the possibility of nonmicrobial methane formation in the mitochondria and eukaryote cells in both plants and animals. The biological role of this process in the mammalian physiology is largely unmapped, but it is conceivable that the formation and constant buildup of reactive oxygen species in the mitochondria are integral parts of a broader system. The initial in vitro studies led to the proposal that a continuous lack of the electron acceptor oxygen will maintain an elevated mitochondrial $NADH/NAD^+$ ratio, causing reductive stress. Electrophilic methyl groups bound to positively-charged nitrogen moieties such as those in phosphatidylcholine molecules may potentially act as substitute electron acceptors as a consequence of which methane may be liberated. Thus, priming during hypoxia occurs as a progressive process involving depressed electron transport, the loss of cytochrome c and antioxidants, and the triggering of methane release

* Corresponding author: Institute of Surgical Research, University of Szeged, Hungary Szőkefalvi-Nagy B. u. 6, 6720 Szeged, Hungary. Email: boros.mihaly@med.u-szeged.hu.

during the abnormal formation of ROS induced by reoxygenation or reperfusion. It is therefore possible that the formation and emission of methane in mammals may be connected with hypoxic events leading to, or associated with a mitochondrial dysfunction. This also points to a role of methane as an alarm signal for the development of mitochondrial responses under hypoxic conditions, and accordingly it may be a biomarker for the diagnosis of such events.

Keywords: methanogenesis, non bacterial, biotic, methane

2.1. Introduction

Methane (CH_4) is a small molecule, the most reduced form of carbon. It plays a distinguished role in tropospheric and stratospheric chemistry, where the bulk of the released CH_4 is oxidized to CO_2 due to its reactivity with hydroxyl radicals (formed by the photoreaction of tropospheric ozone and water vapor) (Cantrell, 1990; Hurkuck, 2012). On a global scale, an approximately 2.5-fold increase has been observed in the atmospheric CH_4 concentration since preindustrial times, and it is therefore regarded as a significant greenhouse gas of growing ecological importance (Denman, 2007; Mitchell, 2013).

CH_4 is also a component of the omnipresent gaseous environment which maintains aerobic, human life on Earth. In this atmosphere, CH_4 is intrinsically nontoxic, biologically inactive and inert. It is a simple asphyxiant, which means that tissue hypoxia might occur when CH_4 displaces air and hence O_2 in a restricted area, and when the concentration of O_2 is reduced to below 18% in the internal milieu of the body. However, the recognition of the importance of other pollutants, including nitric oxide (NO), as signaling mediators has radically altered the view on the roles and functions of gases in physiology. Further discoveries on the generation of carbon monoxide (CO) and hydrogen sulfide (H_2S) by the mammalian cell have deepened knowledge on the significance of endogenous gaseous products, research on "gasotransmitters" and derivatives has been intensified, and this has become a topic of constant scientific interest (Dawson, 1994; Wang, 2003; Ryter, 2004; Wu, 2005; Lamon, 2009; Motterlini, 2010).

It has become clear that gas signals and gaseous mediators form complex intra- and intercellular pathways and regulate numerous physiological processes, separately, or more often in antagonistic or synergistic ways (Gadhia, 2013; Rochette, 2013; Szabo, 2013). In this line, the activity or toxicity of known gas mediators (i.e., NO, CO and H_2S) is related to their tendency to react with biologically important molecules. In contrast, CH_4 is nonreactive, passes through biomembranes, diffuses into the plasma, cytosol or mitochondrial matrix, and is able to target the cell organelles freely. Thanks to these characteristics, it can readily change the symbiosis with other gas molecules in closed spaces. Nevertheless, such in vivo relationships and their outcome currently remain basically unknown, because the detection of the intracellular distribution of these molecules is technically limited and this question has hardly been examined.

2.2. Nonmicrobial CH_4 Generation

Mammalian methanogenesis was at one time widely considered to be an exclusive indicator of gastrointestinal (GI) carbohydrate fermentation by the anaerobic flora. This commonly held notion was challenged, however, when in vitro and in vivo studies and other investigations revealed the possibility of nonmicrobial CH_4 formation in the mitochondria and in eukaryote cells in plants and animals.

As long ago as 2003, it was shown that hypoxia can lead to the generation of measurable amounts of CH_4 nonbacterial origin in isolated rat liver mitochondria (Ghyczy, 2003). When the possible causes were explored, increasingly high amounts of CH_4 were reproducibly generated after the addition of ascorbic acid and hydrogen peroxide (H_2O_2), and its formation was related linearly to the quantity of mitochondria incubated, the amount of H_2O_2 added and the pH of the reaction mixture. Catalase, which decomposes H_2O_2, abolished the increase in CH_4 production, which indicated that mitochondrial H_2O_2 is required for the hypoxic activation of the CH_4-generating reaction. It is important to note that the rate of H_2O_2 formation in rat liver tissue is approximately 10% of the rate of total oxygen consumption, and catalase is absent in the mitochondrial matrix.

These findings remained rather unnoticed until 2006, when Keppler and colleagues provided direct evidence of CH_4 generation in plants under aerobic conditions (Keppler, 2006). This key paper was followed by many plant studies that either supported or disagreed with the initial findings, but aerobic nonbacterial CH_4 release was later clearly identified. More importantly, it has been shown that oxido-reductive stress or even physical injury by cutting elicits aerobic CH_4 emission (Wang, 2011) and many other publications have confirmed the causes of nonmicrobial CH_4 release in cell cultures under various stress conditions (McLeod, 2008; Vigano, 2008; Bruggemann, 2009; Messenger, 2009; Wishkerman, 2011). In a comprehensive literature review, after an evaluation of all the experimental work published on aerobic CH_4 generation in terrestrial plants, and "despite much uncertainty and skepticism", Bruhn and co-workers concluded that the phenomenon was true (Bruhn, 2012). Detailed overviews for those interested in the subject of CH_4 generation in plants were presented by Keppler and colleagues in a 2009 concept article and more recently by Wang and co-workers (Keppler, 2009; Wang, 2013). In this latter excellent review, the emission CH_4 from plants is now clearly linked to a hitherto overlooked defense reaction.

In 2008, further studies were undertaken which demonstrated aerobic CH_4 emission in cultured endothelial cells exposed to hypoxia and metabolic distress. The latter included the inhibition of glucose uptake and anaerobic glycolysis, the application of site-specific inhibitors of the mitochondrial electron transport chain, alone or in combination with glycolysis inhibitors, the application of an uncoupling agent, and treatment of the cells with increasing concentrations of the hydroxyl radical-generating Udenfriend system. These studies provided evidence of stress-induced nonbacterial CH_4 production in eukaryotes. The results revealed that disturbance of the normal mitochondrial function led to significant CH_4 generation in endothelial cells (~ 2-23 nmol/mg), depending on the nature and intensity of the metabolic distress, and a similarly high and dose-dependent level of CH_4 generation was measured after free-radical attack of the Udenfriend reaction.

In parallel with these studies, significant CH_4 release was also demonstrated in whole animals under hypoxic stress conditions (Boros, 1999; Ghyczy, 2008b; Boros, 2012; Tuboly, 2013), where changes in exhaled or released CH_4 changes were detected in vivo. Hypoxia/reoxygenation or ischemia/reperfusion (I/R) initiates a complex biochemical cascade in which the perturbed mitochondrial electron carrier machinery leads to the generation of reactive oxygen species (ROS) and other radicals. Following the hypoxia/reoxygenation cycle, the influx of activated polymorphonuclear leukocytes is accompanied by further ROS formation in the reperfused tissues. These antigen-independent responses interact and amplify each other, finally leading to impaired microhemodynamics, functional and structural cell damage, and remote or systemic inflammatory complications. Indeed, experimental I/R studies are usually conducted in order to analyze ROS-induced reactions and the in vivo effectiveness of anti-inflammatory or antioxidant therapies on tissue integrity and function. In our first pilot study, breath CH_4 was measured in exhaled air during closed-circuit anesthesia and in a further I/R group the animals were pretreated with antibiotics targeting the intestinal CH_4-generating bacterial flora. Reperfusion was associated with a significant increase in breath CH_4, and the concentration of which was similar to that observed in nontreated I/R animals (Boros, 1999). In another large animal model of intestinal vascular occlusion and reperfusion, the results provided further evidence that temporary ischemia leads to significant CH_4 liberation in the exhaled air during the early phase of reoxygenation (Ghyczy, 2008b).

More importantly, it has been shown that the CH_4-producing phenomenon can be mimicked by the administration of sodium azide (NaN_3), a compound known to disrupt electron transport flow, by specifically binding to cytochrome c oxidase (complex IV). The nonmicrobial CH_4 generation was demonstrated when cultured endothelial cells were exposed to site-specific inhibitors (NaN_3 and NaCN) of the electron transport chain (Ghyczy, 2008). This work was followed by another comprehensive study in which CH_4 formation was detected in tobacco, grape vine and sugar beet cell cultures (Wishkerman, 2011). What was particularly interesting was that, under nonstress conditions, the cells produced only small amounts of CH_4, but the production was increased by one to two orders of magnitude when NaN_3 was added to the cultures (Wishkerman, 2011). The main effect of NaN_3 in plants is its direct and irreversible binding to the heme cofactor of cytochrome c oxidase, the final electron acceptor of the electron transport chain; thus, it can also be considered a tool via which to study mitochondrial oxido-reductive stress in plants.

As a further step, the in vivo CH_4 production profile was determined in living animals after the induction of mitochondrial distress by chronic NaN_3 administration. The magnitude of the whole-body CH_4 emission was compared with that in animals treated with antibiotics to eradicate CH_4 producing intestinal bacteria. In the present study, CH_4 exhaled from the airways or discharged through the skin and body orifices was quantified by means of a whole-body CH_4 detection setup (see later), and the emanation of endogenous CH_4 was detected throughout the 8-day experiments (Figure 1). The administration of rifaximin caused a decrease in CH_4 output, but the production significantly exceeded the control and background values. In conclusion, the inhibition of mitochondrial cytochrome c oxidase by chronic NaN_3 administration induces a significant level of CH_4 generation, independently of the methanogenic flora (Tuboly, 2013b). It was therefore proposed that failure of the mitochondrial electron transport chain (METC) could be the initiating event in CH_4 generation (Tuboly, 2013).

Thus, the overall evidence from these findings suggests that the excretion of CH_4 in the breath in mammals reflects not only intestinal bacterial fermentation, but also unidentified nonbacterial generation induced from target cells. In the setups involving NaN_3 administration, the inhibition of complex IV could cause a rearrangement of the electron transport and result in increased ROS production at complexes II and III. It is therefore possible that the formation and constant buildup of ROS in the mitochondria are components of a reaction, which furnishes CH_4 in the living organism. As in plants, the release of CH_4 may be associated with ROS generation after transient intracellular oxygen deprivation (Keppler, 2006; Ghyczy, 2008a; Messenger, 2009; Wishkerman, 2011; Tuboly, 2013b). The results published to date imply that aerobic CH_4 formation that it is highly variable in time and location, and may be an integral feature of cellular responses to changes in oxidative status in all eukaryotes (Table 1).

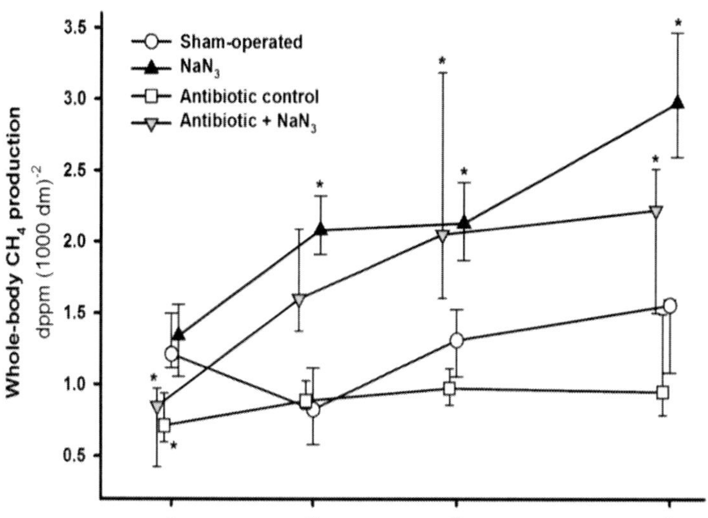

Figure 1. Whole-body CH_4 production in rats on days 1-8 of the investigation after NaN_3 or rifaximin+NaN_3 treatment. During the first measurement, all the rats were untreated, except for the antibiotic-treated group, which had been pretreated for 3 days. The elimination of the intestinal bacteria led to a decrease in CH_4 emission, but it remained measurable, and by day 8 the level was significantly higher in the animals subjected to the NaN_3 challenge than in the control group. Median values and 75th and 25th percentiles are given, * $p < 0.05$ vs the sham-operated, antibiotic-treated control group (adapted from Tuboly, 2013b).

Table 1. Publications on aerobic CH_4 formation and its possible function in animal cells

Reference	Topic
Boros et al., 1999	Approximately 3-fold increase in exhaled CH_4 after intestinal ischemia and reperfusion in dogs.
Ghyczy et al., 2001	Hypothesis of hypoxia-induced endogenous CH_4 formation. Comprehensive overview of pathological states associated with elevated reductive stress from the aspect of possible methyl donor compounds (phosphatidylcholine, S-adenosylmethionine, choline, etc.).
Ghyczy et al., 2003	In vitro measurements of CH_4 generation on isolated mitochondria after hypoxia or treatment with various reducing agents. Model chemical reaction with essential constituents (explained in section 4.), clarification of basic properties of the reaction: exothermic, pH-dependent. Essential role of H_2O_2-derived hydroxyl radicals revealed in the system.
Pimentel et al., 2006	Experimental data suggest that CH_4 slows the small intestinal motility and augments the contractile activity, thereby exerting biological effects on the enteral nervous/smooth muscle system (*subsequent papers on methane-motility relations are omitted from this table for the sake of clarity).
Ghyczy et al., 2008a	Demonstration of CH_4-generating capability of PC metabolites in many cases, proportional to the number of methyl groups in the compound. These substances effectively inhibit ROS formation, by scavenging hydroxyl radicals. Endothelial cell cultures emit measurable amounts of CH_4 after hypoxia, glucose deprivation, cytochrome c oxidase inhibition, uncoupling of the METC and ATP synthesis and free radical attack.
Ghyczy et al., 2008b	Elevation of CH_4 levels in the exhaled air during mesenteric I/R in a canine model. PC has strong anti-inflammatory effects and CH_4 formation is a diagnostic parameter for reduced oxido-reductive stress and inflammatory response.
Keppler et al., 2009	A review paper summarizing results of several research groups. Theories and future perspectives of aerobic, non-bacterial CH_4 formation are presented ('1st Workshop on Aerobic Methane Formation', 2009, Mainz, Germany).
Boros et al., 2012	An exogenous, normoxic CH_4-O_2 mixture exerts anti-inflammatory effects after mesenteric ischemia and reperfusion. Inhalation of a gas mixture containing 2.5% CH_4 for 15 min at the beginning of reperfusion attenuated leukocyte activation and oxidative and nitrosative stress in vivo.
Tuboly et al., 2013b	Long-term administration of NaN_3 induces complex IV inhibition (chemical hypoxia) and elevates excreted CH_4 levels in vivo in rats.
Tuboly et al., 2013a	Technical report on exhaled methane detection, using non-invasive real-time photoacoustic near-infrared laser spectroscopy.

2.3. Bacterial Sources

As described above, nonbacterial generation may occur in vivo, but in the human body CH_4 is produced more conventionally (and perhaps on a larger scale) via bacterial enzymatic metabolism. Nevertheless, the significance of this cooperation is far from clear. In this complex ecosystem, CH_4 is formed in consequence of the gut microbiological activity, whereby CO_2 is reduced to CH_4 by H_2 produced as an electron donor by anaerobic fermentation. CH_4 breath testing using lactulose ingestion reveals two distinct human populations, CH_4 producers and non-producers, production usually being defined as more than a 1 ppm increase above the atmospheric CH_4 concentration (Bond, 1971). The causes and consequences of this pattern are subjects of debate and investigation; nevertheless, it is noteworthy that the proportion of CH_4 producers has remained relatively stable in the Western population during recent decades, despite changes in GI medication and diets (Levitt, 2006).

Methanogenic microorganisms of the GI tract comprise a phylogenetically independent group, the Archae, a group well distinguished from the usual bacteria and the eukaryotes. A clinically important feature is the lack of peptidoglycan in the cell wall, making them susceptible to certain antibiotics. As an obligate anaerobe, this taxon requires a redox potential of less than -300 mV for growth, a condition that is present in the GI tract of mammals and in other anoxic environments (underwater, rice paddies or wetlands). The mammalian methanogens can be divided into two groups: H_2/CO_2- and acetate-consumers, which are members of a microbial consortium, where methanogens obtain substrates from higher levels, from H_2-producers or acetogens (Morris, 2012). Methanogens are compelled to compete with other microorganisms, such as sulphate-reducing bacteria for the common substrates in the human colon (Strocchi, 1994). Although, some H_2/CO_2-consumers are also capable of utilizing formate, acetate is consumed by only a limited number of strains. H_2-consuming methanogens are also important in maintaining low levels of atmospheric H_2 and in producing biogas, a possible sustainable alternative for fossil fuel replacement.

It is currently widely accepted that the bulk of the CH_4 produced by anaerobic fermentation in the mammalian intestine is excreted via the lungs, and breath testing has therefore become a tool for the diagnosis of certain GI conditions. Nevertheless, in consequence of its physicochemical properties, CH_4 is distributed evenly across membrane barriers, traverses the mucosa and enters the splanchnic microcirculation freely. Thus, it should be stressed that the production of CH_4 is reflected not only in the flatus or in the exhaled air, but also in its passage through body surfaces. Indeed a recent study demonstrated the uniform release of CH_4 through the skin in healthy individuals (Nose, 2005).

2.4. Aerobic Chemical Processes Leading to CH_4 Formation

After the key paper of Frank Keppler et al. describing the in situ formation of CH_4 in plants, the accompanying editorial commentary asked what mechanism could be involved in the production of the fully reduced CH_4 gas in an oxidizing environment (Lowe, 2006). Though many publications are now available on the in vivo generation of CH_4 under distinct pathophysiological conditions, the mechanism of aerobic CH_4 generation is basically unexplained, and at present a number of cellular sources are subjects of investigation. This is mainly due to the various possible sources and unknown reaction pathways that can be envisaged. However, as is evident from the following list, the common denominator of the different experimental setups has been hypoxia or ischemia and reperfusion with transient oxido-reductive stress.

- The first hypothesis proposed the possible role of electrophilic methyl (CH_3) group (EMG)-containing biomolecules, leading to CH_4 formation under reductive stress conditions (Ghyczy, 2001). It has been hypothesized that, if the sources of the CH_3 groups are molecules such as phosphatidylcholine or S-adenosylmethionine, the substitution will lead to the generation of CH_4 (Figure 2). Mitochondrial experiments partially confirmed this theory (i.e., under highly reductive conditions, potentially electron acceptor biomolecules can be reduced, leading to the formation of CH_4).
- After the above hypothesis has been put forward, our research group identified CH_4 formation in animal mitochondria (Ghyczy, 2003). To clarify the possible mechanism, an experimental setup was established which involved substances ubiquitously available in biotic systems, but without biological structures. In a chemical model reaction, the formation of CH_4 from choline was demonstrated in the presence of H_2O_2, catalytic iron and ascorbic acid. In this exothermic reaction, CO_2 and CO are formed in parallel with CH_4 generation. The

components of the reaction mixture (other than uncomplexed iron) are known to be present in comparatively high concentrations in biological systems, but the in vivo significance of these chemical processes remained unknown. Thus, during the past decade other possible mechanisms have been proposed, with partial modification of the previous presumptions (Ghyczy, 2003; Ghyczy, 2008a; Keppler, 2008; Messenger, 2009).

- Abiotic, purely chemical reactions at high temperature or elevated pressure are known to lead to CH_4 formation. Such examples include the incomplete burning of biomass (Crutzen, 1990) or the Fischer-Tropsch reaction involving CO and H_2, the latter being extensively used industrially to produce CH_4 gas (Fischer, 2008). However, this reaction is unlikely under ambient conditions.
- Through the use of simple buffers with compounds thought to be involved in aerobic CH_4 formation, a ROS-generating Udenfriend system was established, which consisted of an EMG or methyl-donating compound (MDC) / electron acceptor (e.g., choline chloride, betaine or phosphatidylcholine), reducing agents (e.g., ascorbate, NADH, NADPH, dithiothreitol or N-acetyl-L-cysteine), H_2O_2 and uncomplexed Fe^{3+} as catalyst. In this series of experiments, when the CH_4-generating capacity of choline metabolites was tested, the amount of CH_4 formed was generally observed to increase linearly with the number of CH_3 groups in the molecule (choline > 2,2-dimethylethanolamine > 2-methylethanolamine > ethanolamine) (Figure 3).
- Althoff and co-workers proposed an alternative but similar approach under more controlled conditions (Althoff, 2010) for the aerobic formation of CH_4 by the oxidation of ascorbic acid with iron compounds and H_2O_2. In contrast with our findings, they reported that CH_4 generation also occurs in the absence of choline or another MDC.
- On the basis of our experimental results with MDCs, Althoff et al. tested methionine and methionine sulfoxide, which exhibited the highest CH_4 formation ability. Their findings are in accordance with the concept of EMGs, which also explains the differences in CH_4 formation ability between similar molecules. For example, although choline readily produces methane, the analogous molecule betaine, which also contains an EMG, does not. In betaine, the nitrogen-bounded CH_3 group is sterically shielded and electrons are repelled by the electronegative N atom, reducing its reactivity. Moreover, the

sulphur-bounded CH_3 group in methionine was unambiguously identified by stable isotope labelling techniques as the origin of the C atom of the CH_4 molecule (Althoff, 2012).

Figure 2. Proposed mechanism of CH_4 formation from biomolecules containing an electron-deficient CH_3 group. Some molecules containing a CH_3 group bound to a positively charged N or S atom are able to accept electrons from NADH, thereby oxidizing NADH to NAD^+ in the absence of oxygen, or, more generally, in conditions of reductive stress (adapted from Ghyczy, 2001).

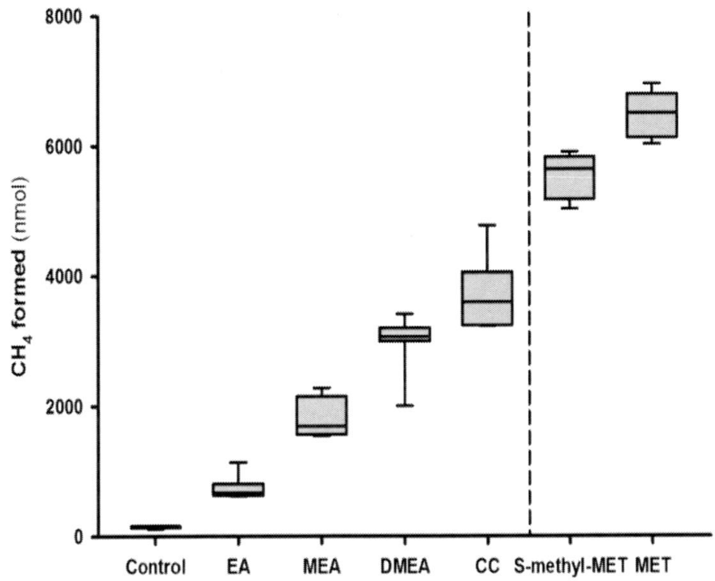

Figure 3. CH_4-producing capacities of choline metabolites in the hydroxyl radical-generating Udenfriend system (see explanation in the text). There is a linear correlation between the number of CH_3 groups in the molecule and the amount of CH_4 formed. Methionine and S-methylmethionine generate more CH_4 than do choline metabolites. EA: ethanolamine, MEA: 2-methylethanolamine, DMEA: 2,2-dimethylethanolamine, CC: choline chloride, S-methyl-MET: S-methylmethionine, MET: methionine. Median values and 75th and 25th percentiles are given (adapted partly from Ghyczy 2008a).

In these model experiments, the highest level of CH_4 formation was measured at around pH 3.0 under hypoxic conditions, which can be explained at least partially by chemical reasons. Deprotonation of ascorbate (pK_a=4.37) leads to acidic pH, at which the production of hydroxyl radicals is favored unlike at higher pH values, where H_2O_2 produces nonreactive oxygen species (Kremer, 2003; Althoff, 2010). A more thorough analysis of postulated reaction routes and detected and possible intermediates is to be found in Althoff's referred works (Althoff, 2012).

- A very similar process was described in plants in 2006 by Keppler and co-workers, and subsequently by other research groups (Keppler, 2006; Keppler, 2008; McLeod, 2008). CH_4 release was reported from both fresh and dried foliage, in some cases after UV light irradiation. Later, pectin was shown to be the one of the origins of CH_4, the carbon atom being that of the methoxy group. We refer here to an excellent review (Wang, 2013a), in which the authors summarize the key characteristics of possible CH_4-generating compounds, and in particular those who involved in methanogenesis in plant cells.

It seems that various organic compounds can serve as the source of the liberated CH_4. The common property of these molecules is the presence of a CH_3 or related group: methoxy (Keppler, 2008; Vigano, 2008), acetyl (Messenger, 2009) or hydroxymethyl (Vigano, 2009). Practically ubiquitous compounds such as pectin, lignin, lipids or fatty acids may serve as depots of these functional groups. Another common feature of the chemical processes explored so far is the presence of a highly reactive radical. Hydroxyl radicals can be produced in the iron-catalyzed Fenton or Haber-Weiss reactions during hypoxia and reoxygenation. It should be added that for a long time the biological relevance of hydroxyl radical formation was considered limited, simply because of its high reactivity toward virtually every biomolecule. Likewise, it was proposed that the hydroxyl radical encounter too many irrelevant biological targets within a short diffusion distance. There is no known detoxification system for the hydroxyl radical in the human body, and its generation is therefore a critical process (Sheu, 2006). Interestingly, it has been shown that H_2 can alleviate hydroxyl radical-induced cytotoxicity in living cells (Ohsawa, 2007).

By virtue of its in situ reactivity, the hydroxyl radical is able to break intramolecular bonds in membranes, leading to the formation of CH_4.

However, the expanding literature on aerobic CH_4 generation has not gone hand in hand with an understanding of the complexities of intracellular redox chemistry. It is important to point out that the previously presented chemical processes are feasible, but still hypothetical modes of CH_4 release in biotic, aerobic environments. Nevertheless, if we consider the whole scale of the published in vitro findings, basically two postulations can be made.

1. A prerequisite of CH_4 formation under ambient conditions is the availability of a donor compound containing one or more EMGs or other related functional groups, and
2. a further prerequisite is the presence of oxidative or reductive stress (elevated reducing potential and radical reactions; see later).

No direct enzymatic route of aerobic, nonbacterial biotic CH_4 generation has yet been proven. The positive correlation between CH_4 production and temperature (30-70 °C), the possibility of CH_4 liberation from dried leaves (Keppler, 2006) and the heavily exothermic nature of the model reaction (up to 95 °C) (Ghyczy, 2003) suggest a chemical explanation rather than enzymatic catalytic process (Figure 4).

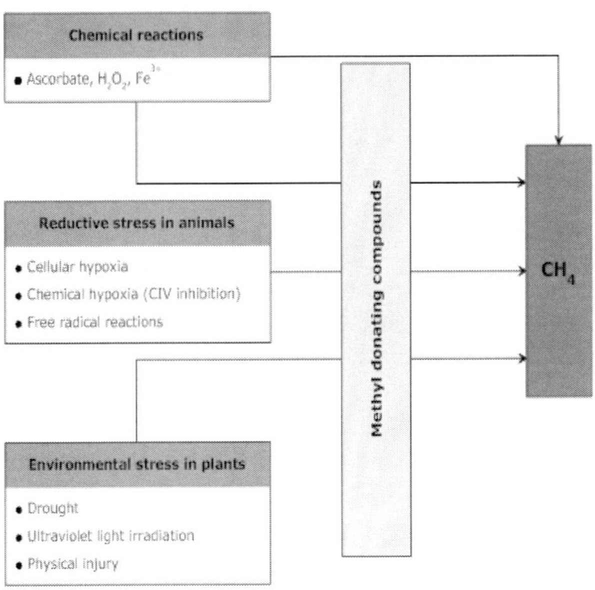

Figure 4. Possible routes for aerobic, non-microbial methane formation.

2.5. The Mitochondrial Electron Transport Chain and the Oxido-Reductive Balance

As either substrates or metabolites, gases are crucial components of the aerobic homeostasis, and the redox balance is an essential feature of aerobic life. All aerobic organisms have evolved a range of endogenous mechanisms to allow them to utilize O_2 without succumbing to its inherent toxicity. Another major threat to this controlled equilibrium is hypoxia, as the absence of electron acceptor oxygen leads to progressive functional and structural cell damage. In the mitochondria, substrate oxidation by the METC creates a proton gradient across the inner membrane, which becomes a driving force for ATP synthesis. For optimal performance, the electrons shuttle must be in a steady state as regards the two entry points and the final utilization points, and oxygen availability is therefore crucial. Hypoxia entails a shift in reducing potential to a higher than normal reducing power, which may result in a depression of mitochondrial ATP formation. Re-establishment of the O_2 supply, however, is also precarious, as the disturbed intracellular biochemistry may lead to the generation of excess superoxide, which forms the basis of “oxidative stress”. This term is applied to in vivo situations in which damage is caused by elevated levels of ROS or other free radicals, either directly or indirectly. The targets of ROS may include proteins, sugars and structure-forming phospholipids, and it is widely accepted that several disease states are linked to oxidative stress, even though the potential of ‘antioxidant’ compounds to prevent or cure ROS-induced processes is very limited (Gutteridge, 1999). More importantly, the precise mechanisms by which ROS produce cellular injury are often elusive.

The METC is a potential source of ROS, even under baseline conditions, where both complexes I and III are probably the main production sites. Nevertheless, it has been shown that approximately 1-2% of the O_2 consumed during normal physiological respiration is converted into superoxide radicals, due to the natural leak of electrons from the electron transport chain (Nohl, 1978). ROS also play a regulatory role in cellular metabolic processes by activating of various enzymatic cascades and several transcriptional factors (Corbi, 2013). As ROS are produced at significant rates by reactions intrinsic to the normal aerobic metabolism, “redox homeostasis” appropriately includes an intracellular balance between ROS generation and scavenging (Droge, 2006).

In summary, a steady state of reducing power or redox balance is as important for the normal functioning of aerobic cells as is a constant pH. Attention has been focused in the past on "oxidative stress", which has long been assumed to be the main cause of ROS activity and its damaging consequences in biological systems. Nevertheless, the reverse imbalance, "reductive stress", can often be the underlying cause of the pathological conditions which are today named oxidative stress. Reductive stress can be normalized only by electron acceptors, which can include EMGs. In this way, we hypothesized that a mechanism which may operate in biological systems during reductive stress may be the capture of electrons by EMGs and the consequent irreversible evolution of gaseous CH_4. It was proposed that a continuous lack of the electron acceptor O_2 will maintain an elevated mitochondrial NADH/NAD^+ ratio, causing reductive stress and the formation of a nucleophilic hydride ion, which may be transferred to the EMG (Ghyczy, 2001). It is likely that the same mechanisms occur in animal and plant mitochondria, and indeed it has been demonstrated that alternative electron acceptors in plants may help to prevent the overreduction of electron carriers (reductive stress) in fluctuating light or under light-induced stress conditions. Thus, it has been suggested that the priming of CH_4 formation commences during ischemia (1) as depressed electron transport in the inner mitochondrial membrane is accompanied by the loss of cytochrome c and antioxidants; and (2) the triggering of CH_4 release at reperfusion determined by the interplay of mitochondrial membrane potential with mitochondrial matrix ROS production (Tuboly, 2013b).

2.6. A Theory of a Membrane-Associated Mechanism for CH_4 Formation

A continuing lack of O_2 during ischemia or other pathologies involving the interruption of the electron flow will lead to an anomalously increased reducing power and dissipation of the mitochondrial transmembrane potential. Mitochondria are not only targets, but also sources of oxido-reductive stress, and mitochondrial ROS generation is implicated in a wide variety of diseases, acute or chronic stress and I/R injuries (Boland, 2013; Dikalov, 2013; Mukherjee, 2013; Ngo, 2013; Wang, 2013a; Huang, 2014). A number of conditions lead to impairment of the redox homeostasis, thereby inducing the formation of ROS by transferring the excess electrons to oxygen outside the transport chain, at a site where membrane-bound phosphatidylcholine or free choline is present. It must be emphasized that choline may also be liberated from phosphatidylcholine through the redox-sensitive activation of phospholipase D. This enzyme hydrolyzes phosphatidylcholine to phosphatidic acid and choline, and has been implicated as a signal-activated key enzyme in a wide range of physiological responses (Tappia, 2006). Choline belongs among the primordial molecules from which the first cells were built up in terrestrial life. In the course of evolution, its derivatives have acquired several functions, including the formation of membrane barriers between the living and the nonliving worlds.

The mechanistic role of hypoxic, nonmicrobial CH_4 generation remains to be established, but one of the possible explanations could be an endogenous need for the sparing or regeneration of membranes. Complex IV inhibition causes ROS production and serious membrane loss due to ROS-induced lipid peroxidation (Musatov, 2012). Peroxidation is an immediate chain reaction which causes a fundamental breakdown of biomembranes, leading to decompartmentalization, loss of integrity and cell death (Halliwell, 1993). As a consequence, membrane sparing and recovery are particularly important in such an environment. To avoid the potentially fatal outcome of an increased oxido-reductive power, molecular participants of the living system should be quickly brought into use to regenerate membranes, essential for the maintenance of the cellular homeostasis via channels, pores and membrane proteins. Theoretically, molecules with membrane regenerative potential for phospholipid bilayers might be re-utilized. It is possible that, during such a process, compounds rich in ethyl (C_2H_5) and CH_3 groups can be reduced by electron acceptance and then used to seal or build up membranes. This process is accompanied by the generation of fully reduced CH_4 (Ghyczy, 2001). Accordingly, we propose that CH_4 is the end-product of a protective mechanism linked to membrane defense or regeneration during ROS-induced damage. Furthermore, the increased uptake of phosphatidylcholine exerts an anti-inflammatory influence in various experimental scenarios (Hartmann, 2009; Kovacs, 2012), and we have previously shown that pretreatment with phosphatidylcholine reduces the exhaled CH_4 concentration during intestinal I/R injury (Ghyczy, 2008b).

To reiterate, it is possible that CH_4 accumulation may change the symbiosis with other gas molecules in the intracellular spaces and this can influence enzymatic reactions. Inhaled CH_4 will move from the alveoli into the circulation, throughout which it is distributed rapidly and evenly. If there are no physical barriers to prevent cellular entry, CH_4 may be taken up by virtually all cells in the organism. Thereafter, the concentration in all regions should be equal to the equilibrium concentration in the atmosphere (where it is normally 1-2 ppb), or to that in the inhaled air or to the concentration within the lumen of the GI tract, if these are the sole or main sources of CH_4. Indeed, the CH_4 concentration in the breath is usually greater than 1 ppm in only 30-60% of humans, but a recent study has shown the release of a uniform ~150 pg cm^{-2} CH_4 through the skin in healthy individuals (Nose, 2005). This suggests that CH_4 transported by the circulating blood is excreted by the lungs only when a certain threshold is reached. The fate of intracellular CH_4 is an open question, but there are many hydrophobic and hydrophilic interfaces in the cytoplasm and, as a small nonpolar molecule, CH_4 may enter the hydrophobic nonpolar lipid tails of the phospholipid membranes (Serra, 2006). This effect will be even stronger at high salt concentrations, because the hydrophobic interactions are enhanced as a result of the salting-out effect (Hemmingsen, 1985; Docherty, 2007). This entry should be temporary, however, because, without a new supply, CH_4 will enter the circulation and will be excreted through the lungs if the partial pressure is higher than that in the atmosphere. It follows that CH_4 may accumulate transiently at cell membrane interfaces, thereby changing the physico-chemical properties or the in situ functionality of proteins embedded within this environment, and in this case it is able to influence the function of membrane-bound enzymes (Batliwala, 1995).

2.7. Effects of CH_4 on Organ Systems

The oxidative or reductive by-product formation of the cell is central to the resultant physiological responses. Although the generation of nonbacterial CH_4 has been illustrated in detail in various tissues under different conditions, there is still reasonable skepticism as to the physiological importance of the reaction. Moreover, the significance of intestinal CH_4 production in the human physiology and the consequences of endogenous CH_4 generation in the GI system are still open questions. Once released, CH_4 is widely considered to be biologically inactive. However, some data do suggest an association with the small bowel motility as the whole gut transit time is longer in human subjects who produce more bacterial (or nonbacterial) CH_4 than in those defined as nonproducers (Jahng, 2012). More directly, diarrheal conditions such as inflammatory bowel disease (IBD) are negatively associated with CH_4 production, and there is evidence that the production of CH_4 as determined by breath testing is associated with delayed intestinal transit and constipation

(Lee, 2013; Triantafyllou, 2014). Using a physiological model of peristalsis in the guinea pig ileum, Pimentel and co-workers explored the effects of CH_4 on the neuromuscular function (Pimentel, 2006). Gassing the bath with CH_4 significantly increased the amplitude of contraction both orally and aborally in response to a stimulus. In a similar setup, luminal CH_4 infusion reduced the transit time in dogs, and it was hypothesized that CH_4 activates a reflex pathway that leads to slowing in the proximal intestinal segment. Finally, it was concluded that CH_4 acts as a neuromuscular transmitter, resulting in reduced propagation of the peristaltic movement in the intestine.

Nevertheless, the physiological importance of endogenous CH_4 to other systems in the human body has not yet been evaluated. Interestingly, a historical paper reported an increased survival time in hemorrhaged rats after treatment with a CH_4-air mixture. Information on the cardiovascular roles of CH_4 is sparse, but likely to expand significantly in the future, following the demonstration that CH_4 protects the tissues by mitigating the effects of oxidative stress following an I/R insult (Boros, 2012). In this animal model, normoxic ventilation with 2.5% CH_4 was used, and no side-effects on the blood-gas chemistry were observed under baseline conditions. Similarly, the inhalation of CH_4 had no influence on the macrohemodynamics in unstressed animals, while at this concentration CH_4 supplementation significantly diminished many of the damaging consequences of hypoxia and reoxygenation. The levels of tissue ROS generation were reduced, the mesenteric vascular resistance changes were only moderate, and the intestinal pCO_2 gap (a marker of the microcirculation) tended to normalize after reperfusion. As decreased tissue and plasma granulocyte activities were also found, the effects of CH_4 on the polymorphonuclear (PMN) leukocyte functions were further investigated by using isolated cells. The in vitro results substantiated the in vivo findings, and established that CH_4 exposure specifically decreases the ROS production of activated PMN leukocytes in a hitherto unrecognized reaction pathway (Boros, 2012).

Recent experiments have revealed another in vivo phenomenon possibly related to the effects of CH_4 in biological systems. In these studies, the reperfusion-induced epithelial hyperpermeability was significantly alleviated by an exogenous, normoxic CH_4-O_2 mixture (Figure 5). This finding is probably related to the anti-inflammatory potential of inhaled exogenous CH_4 (Boros, 2012) and the fact that phospholipids and membrane components (choline and its metabolites) are possible sources of endogenous CH_4 formation (Ghyczy, 2008a). These results also imply that not only the origin,

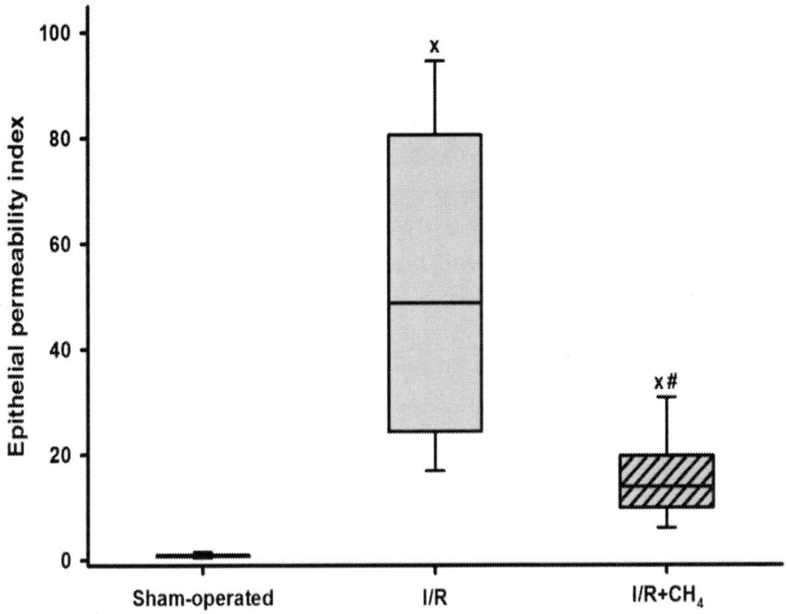

Figure 5. Small intestinal epithelial permeability changes after 45-min mesenteric ischemia followed by 40-min reperfusion in sham-operated, ischemia/reperfusion (I/R) and I/R + CH_4 treated groups. Median values and 25th and 75th percentiles are given. $p<0.05$ was considered statistically significant. [x] $p<0.05$ vs. sham, [#] $p<0.05$ vs. I/R (Meszaros et al., unpublished).

but perhaps also some of the functions of CH_4 are related to biomembranes. The nonpolar CH_4 molecules may dissolve in the cell membrane, therefore influencing its physico-chemical condition, which is essential for the normal functioning of embedded proteins, ion channels and receptors (Phelan, 1991).

The modulator effects of exogenous CH_4 on the production of other gas transmitters are the subject of further research, but in a recent study a significant inhibitory potential was observed on nitrotyrosine formation, which might suggest a link to NO chemistry. As discussed earlier, nonbacterial CH_4 production can be induced in eukaryotes by inhibiting the activity of the mitochondrial cytochrome c oxidase through NaN_3 administration. Interestingly, NaN_3 is also capable of activating the cGMP/PKG/VASP pathway, which is connected with endogenous NO synthesis (Russo, 2008). It has further been demonstrated that, in the presence of H_2O_2 and catalase, NaN_3 can be oxidized, and this might also lead to NO formation (Van Uffelen, 1998). Amongst its multiple regulatory functions, NO controls cell respiration

(Carr, 1990; Ignarro, 1999), directly and rapidly interacts with cytochrome c oxidase, thereby competing with O_2 until free NO is available (Moncada, 2002; Sarti, 2003). In principle, the diffusion-limited reaction between superoxide and NO leads to the formation of peroxynitrite and nitrotyrosine, the latter widely recognized as a biochemical marker of the post-translational modification of proteins. Nitrotyrosine generation involves a covalent modification, but it may also be a dynamic and reversible process (Koeck, 2004). In our in vivo animal model, nitrotyrosine formation was significantly suppressed by increasing the CH_4 input prior to stress induction, which indicates the removal of peroxynitrite from the reaction (Figure 6). This further suggests that the effect of an increased CH_4 input on tissue nitrosative stress is linked to a process which depends on the tissue superoxide and/or NO generation.

2.8. CH_4-Based Diagnostics

Biomarkers are useful tools through which to gain an *understanding of* human or animal *pathophysiology* and the *responses* to therapy, and in this sense the significance of biotic CH_4 production is receiving increased attention. Although CH_4 can be measured in exhaled breath, blood and other tissues, no reliable method earlier existed to determine the dynamics of its in vivo formation. In clinical practice, breath gas analysis has been widely used for the screening of patients with IBD, where the observation of CH_4 concentration changes relative to other gaseous compounds can promote the diagnosis (de Lacy Costello, 2013; Lee, 2013). CH_4 levels are mostly measured by means of gas chromatography (GC), GC-mass spectrometry or He-Ne laser-based technology (Levitt, 2006; Ligor, 2008), but these traditional methods have major technical limitations and the risk of possible artifacts may be significant (Yu, 2004). In our recent studies we have employed near-infrared diode laser-based photoacoustic spectroscopy (PAS), which has proven its relevance in life science applications (Cristescu, 2008). Despite the fact that high-sensitivity PAS-based sensors have proved to be appropriate for multi-component analyses of widely varying gas compositions indicating the potential for complex gas analysis, their use for breath analysis is uncommon (de Lacy Costello, 2013).

The detected CH_4 changes were reproducible; the PAS-based CH_4 sensor is highly specific for CH_4 and has a wide dynamic range, from levels of a few ppm to several thousand ppm and the CH_4 measurements correlate closely with GC measurements. As a further step, rodent (rat and mouse) chambers were built which allow air sampling and the continuous detection of in vivo CH_4 generation. This setup can be used in experiments which correctly reflect the overall, whole-body profile of CH_4 generation (Figure 7).

This system allows in-situ measurements, has low maintenance costs and does not require major technical skills. It operates without chemicals, and the gas-sampling procedure does not demand the use of disposable bags or syringes. The possible inaccuracies associated with adsorption-desorption processes are minimized by using inert materials. The sampling chamber and the flow rate of the gas samples can be optimized for different experiments and the range of application can easily be expanded by the addition of another light source to the system. Furthermore, the instrumentation can be used in acute or chronic experiments without exposing the animals to surgery,

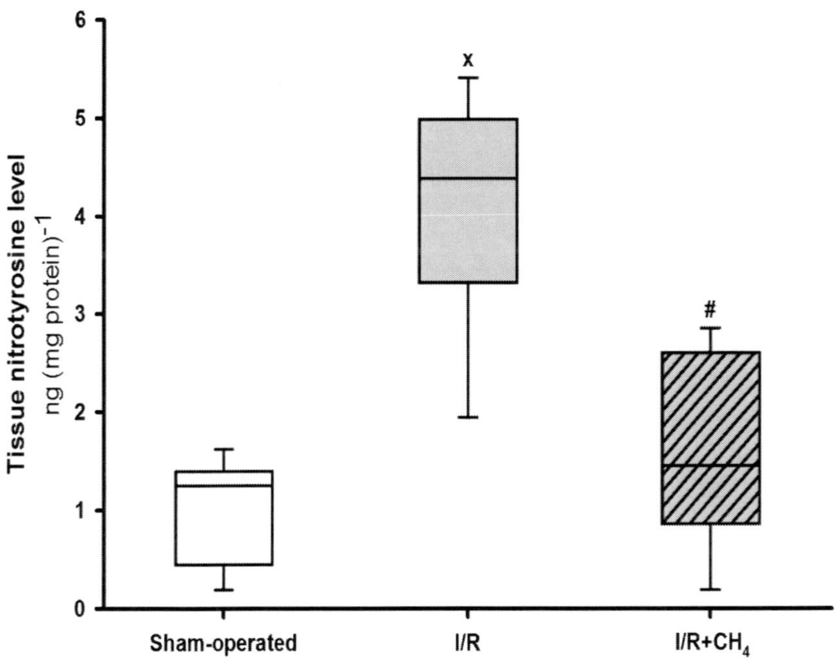

Figure 6. Small intestinal tissue nitrotyrosine levels 15 min after reperfusion in sham-operated, ischemia/reperfusion (I/R) and I/R + CH_4 treated dogs. Median values and 25th and 75th percentiles are given. $p<0.05$ was considered statistically significant. [x] $p<0.05$ vs. sham, [#] $p<0.05$ vs I/R (adapted from Boros, 2012).

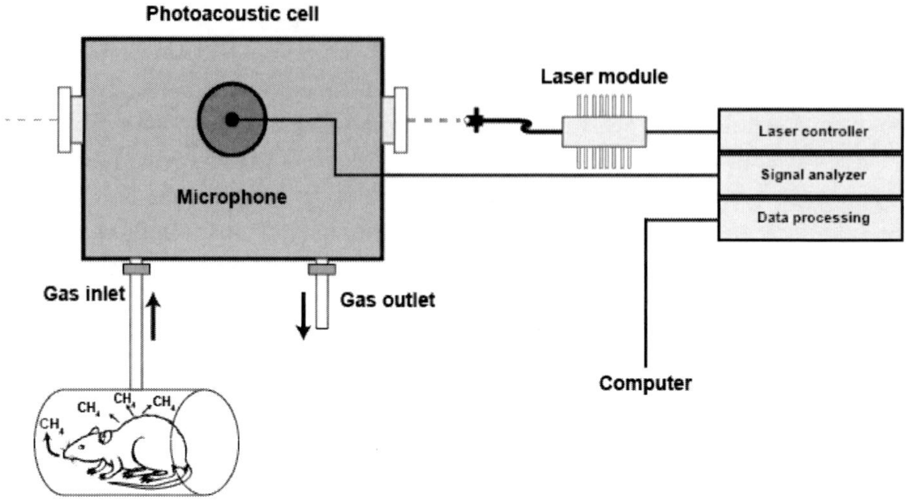

Figure 7. Schematic presentation of the photoacoustic detection setup. Gas samples are drawn by a membrane pump into the photoacoustic cell, where the signal is generated by a diode laser and detected by a microphone. The gas flow rate can be adjusted with a rotameter. The electronic unit provides system control and data processing, and a computer presents data graphically. Arrows indicate the gas flow direction (adapted from Tuboly, 2013a).

anesthesia or other trauma during the measurements. We have additionally demonstrated its facile and reliable usage in human studies on healthy male and female volunteers (Tuboly, 2013a). In a further study the on-line dynamics of alcohol-induced CH_4 generation changes were detected in non-CH_4 producer volunteers, whose negative status was confirmed by continuous PAS-based breath CH_4 analysis prior to ethanol consumption (Tuboly et al., unpublished data, Figure 8). Surprisingly, a 0.291 to 0.375 ppm elevation was observed after an intake of 250 ml ethanol/per day, regardless of the body weight. It should be added that the ethanol metabolism modifies the cellular redox state of the liver mitochondria during chronic alcohol consumption, by increasing the $NADH/NAD^+$ ratio (Zakhari, 2013), and these preliminary results may therefore also suggest a certain role of reductive stress in CH_4 generation.

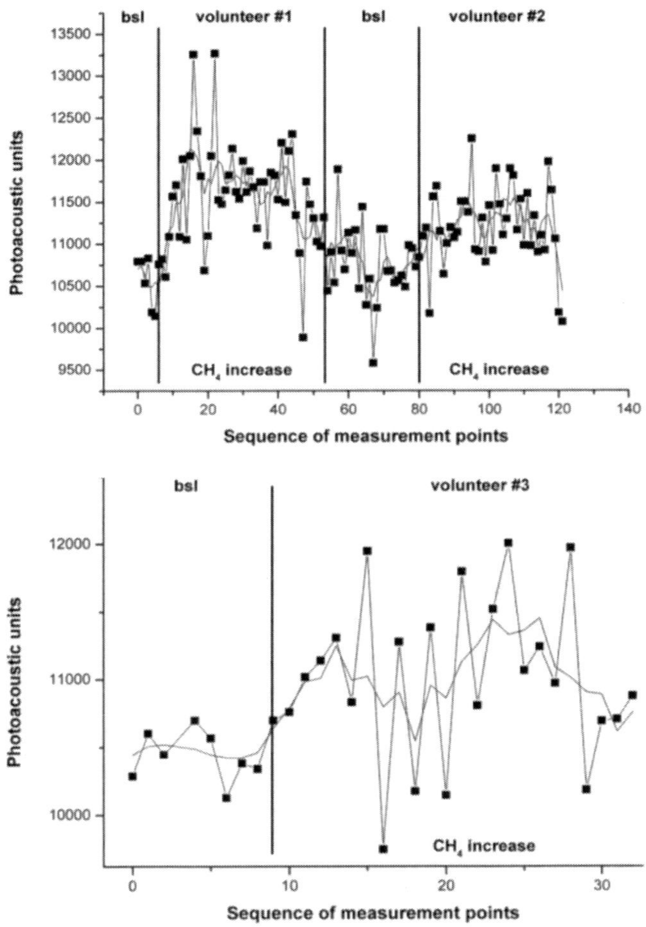

Figure 8. CH_4 production induced by ethanol consumption for 5 days in non-CH_4 producer healthy volunteers (defined as subjects who display a 0 or less than 0.1 ppm elevation in the exhaled CH_4 concentration relative to the CH_4 concentration of the ambient air). An approximately 2-3-fold elevation was observed on PAS-based breath CH_4 analysis after the intake of 250 ml ethanol/day, regardless of the body weight. (1 ppm= 2790 Photoacoustic units) (Tuboly et al., unpublished).

CONCLUSION

In this chapter, we have reviewed the nature of alternative, nonbacterial aerobic CH_4 production, particularly in association with a hypoxia-induced mitochondrial dysfunction. Moreover, we have presented data on recent advances that support the notion that CH_4 may play a role in cellular metabolism. In a consideration of the generation of CH_4, we have emphasized the possible origin of the emission and focused on plausible underlying mechanisms, based on published in vitro and in vivo research findings. Overall, these conclusions demand a consideration of two important aspects of the cellular response to hypoxia: the need for CH_3 groups for CH_4 formation, and a radical reaction for the induction of this response. In this respect, the available information suggests that hypoxia-induced CH_4 generation may be a necessary phenomenon of aerobic life, and perhaps a surviving evolutionary trait in plants and animals. Various data suggest that the excretion of CH_4 in the breath of mammals may predominantly reflect intestinal bacterial fermentation and a variable amount possibly linked to a mitochondrial dysfunction. In this case, the determination of endogenous CH_4 may be a possible indicator or alarm signal of a sustained oxido-reductive imbalance, and CH_4-detecting techniques may be useful tools for the diagnosis of such events. The immediate challenge is to test the proposed hypotheses for aerobic CH_4 formation from different biomolecules and cellular/tissue structures, and to construct a comprehensive picture of their specific importance in animals and humans. The observation that CH_4 formation under aerobic conditions occurs in many ischemic systems has opened up new, interesting and challenging avenues for future research.

Acknowledgments

ET and AM contributed equally to this work. The study was supported by OTKA K104656 and TÁMOP-4.2.2A-11/1/KONV-2012-0073, TAMOP-4.2.2A-11/1-KONV-2012-0035, and co-financed by the European Social Fund in the framework of TÁMOP-4.2.4.A/2-11/1-2012-0001 'National Excellence Program'.

References

Althoff, F., Jugold, A. & Keppler, F. (2010). Methane formation by oxidation of ascorbic acid using iron minerals and hydrogen peroxide. *Chemosphere*, *80*(3), 286-292.

Althoff, F. (2012). Sources and pathways of methane formed in oxidative environments. *PhD Dissertation*. Mainz. Johannes Gutenberg-Universität.

Batliwala, H., Somasundaram, T., Uzgiris, E. E. & Makowski, L. (1995). Methane-induced haemolysis of human erythrocytes. *Biochem J*, *307* (Pt 2), 433-438.

Boland, M. L., Chourasia, A. H. & Macleod, K. F. (2013). Mitochondrial Dysfunction in Cancer. *Front Oncol*, *3*, 292.

Bond, J. H. Jr., Engel, R. R. & Levitt, M. D. (1971). Factors influencing pulmonary methane excretion in man. An indirect method of studying the in situ metabolism of the methane-producing colonic bacteria. *J Exp Med*, *133*(3), 572-588.

Boros, M., Ghyczy, M., Erces, D., Varga, G., Tokes, T., Kupai, K., Torday, C. & Kaszaki, J. (2012). The anti-inflammatory effects of methane. *Crit Care Med*, *40*(4), 1269-1278.

Boros, M., Wolfárd, A. & Ghyczy, M. (1999). In vivo evidence of reductive stress-induced methane production. *Shock*, *12*(6), 56.

Bruggemann, N., Meier, R., Steigner, D., Zimmer, I., Louis, S. & Schnitzler, J. P. (2009). Nonmicrobial aerobic methane emission from poplar shoot cultures under low-light conditions. *New Phytol*, *182*(4), 912-918.

Bruhn, D., Moller, I. M., Mikkelsen, T. N. & Ambus, P. (2012). Terrestrial plant methane production and emission. *Physiol Plant*, *144*(3), 201-209.

Cantrell, C. A., Shetter, R. E., McDaniel, A. H., Calvert, J. G., Davidson, J. A., Lowe, D. C., Tyler, S. C., Cicerone, R. J. & Greenberg, J. P. (1990). Carbon kinetic isotope effect in the oxidation of methane by the hydroxyl radical. *Journal of Geophysical Research: Atmospheres*, *95*(D13), 22455-22462.

Carr, G. J. & Ferguson, S. J. (1990). Nitric oxide formed by nitrite reductase of Paracoccus denitrificans is sufficiently stable to inhibit cytochrome oxidase activity and is reduced by its reductase under aerobic conditions. *Biochim Biophys Acta*, *1017*(1), 57-62.

Corbi, G., Conti, V., Russomanno, G., Longobardi, G., Furgi, G., Filippelli, A. & Ferrara, N. (2013). Adrenergic signaling and oxidative stress: a role for sirtuins?, *Front Physiol*, *4*, 324.

Cristescu, S. M., Persijn, S. T., te Lintel Hekkert, S. & Harren, F. J. M. (2008). Laser-based systems for trace gas detection in life sciences. *Applied Physics B*, *92*(3), 343-349.

Crutzen, P. J. & Andreae, M. O. (1990). Biomass burning in the tropics: impact on atmospheric chemistry and biogeochemical cycles. *Science*, *250*(4988), 1669-1678.

Dawson, T. M. & Snyder, S. H. (1994). Gases as biological messengers: nitric oxide and carbon monoxide in the brain. *J Neurosci*, *14*(9), 5147-5159.

de Lacy Costello, B. P., Ledochowski, M. & Ratcliffe, N. M. (2013). The importance of methane breath testing: a review. *J Breath Res*, *7*(2), 024001.

Denman, S. E., Tomkins, N. W. & McSweeney, C. S. (2007). Quantitation and diversity analysis of ruminal methanogenic populations in response to the antimethanogenic compound bromochloromethane. *FEMS Microbiol Ecol*, *62*(3), 313-322.

Dikalov, S. I. & Ungvari, Z. (2013). Role of mitochondrial oxidative stress in hypertension. *Am J Physiol Heart Circ Physiol*, *305*(10), H1417-1427.

Docherty, H., Galindo, A., Sanz, E. & Vega, C. (2007). Investigation of the salting out of methane from aqueous electrolyte solutions using computer simulations. *J Phys Chem B*, *111*(30), 8993-9000.

Droge, W., Kinscherf, R., Hildebrandt, W. & Schmitt, T. (2006). The deficit in low molecular weight thiols as a target for antiageing therapy. *Curr Drug Targets*, *7*(11), 1505-1512.

Fischer, H., Behrens, M., Bock, M., Richter, U., Schmitt, J., Loulergue, L., Chappellaz, J., Spahni, R., Blunier, T., Leuenberger, M. & Stocker, T. F. (2008). Changing boreal methane sources and constant biomass burning during the last termination. *Nature*, *452*(7189), 864-867.

Gadhia, M. M., Cutter, G. R., Abman, S. H. & Kinsella, J. P. (2013). Effects of early inhaled nitric oxide therapy and vitamin A supplementation on the risk for bronchopulmonary dysplasia in premature newborns with respiratory failure. *J Pediatr* 64(4), 744-748.

Ghyczy, M. & Boros, M. (2001). Electrophilic methyl groups present in the diet ameliorate pathological states induced by reductive and oxidative stress: a hypothesis. *Br J Nutr*, *85*(4), 409-414.

Ghyczy, M., Torday, C. & Boros, M. (2003). Simultaneous generation of methane, carbon dioxide, and carbon monoxide from choline and ascorbic acid: a defensive mechanism against reductive stress? *FASEB J*, *17*(9), 1124-1126.

Ghyczy, M., Torday, C., Kaszaki, J., Szabo, A., Czobel, M. & Boros, M. (2008a). Hypoxia-induced generation of methane in mitochondria and eukaryotic cells: an alternative approach to methanogenesis. *Cell Physiol Biochem*, *21*(1-3), 251-258.

Ghyczy, M., Torday, C., Kaszaki, J., Szabo, A., Czobel, M. & Boros, M. (2008b). Oral phosphatidylcholine pretreatment decreases ischemia-reperfusion-induced methane generation and the inflammatory response in the small intestine. *Shock*, *30*(5), 596-602.

Gutteridge, J. M. (1999). Does redox regulation of cell function explain why antioxidants perform so poorly as therapeutic agents? *Redox Rep*, *4*(3), 129-131.

Halliwell, B. & Chirico, S. (1993). Lipid peroxidation: its mechanism, measurement, and significance. *Am J Clin Nutr*, *57*(5 Suppl), 715S-724S; discussion 724S-725S.

Hartmann, P., Szabo, A., Eros, G., Gurabi, D., Horvath, G., Nemeth, I., Ghyczy, M. & Boros, M. (2009). Anti-inflammatory effects of phosphatidylcholine in neutrophil leukocyte-dependent acute arthritis in rats. *Eur J Pharmacol*, *622*(1-3), 58-64.

Hemmingsen, B. B., Steinberg, N. A. & Hemmingsen, E. A. (1985). Intracellular gas supersaturation tolerances of erythrocytes and resealed ghosts. *Biophysical Journal*, *47*(4), 491-496.

Huang, S. H., Hsu, M. H., Hsu, S. C., Yang, J. S., Huang, W. W., Huang, A. C., Hsiao, Y. P., Yu, C. C. & Chung, J. G. (2014). Phenethyl isothiocyanate triggers apoptosis in human malignant melanoma A375.S2 cells through reactive oxygen species and the mitochondria-dependent pathways. *Hum Exp Toxicol*, *33*(3), 270-283.

Hurkuck, M., Althoff, F., Jungkunst, H. F., Jugold, A. & Keppler, F. (2012). Release of methane from aerobic soil: an indication of a novel chemical natural process? *Chemosphere*, *86*(6), 684-689.

Ignarro, L. J., Cirino, G., Casini, A. & Napoli, C. (1999). Nitric oxide as a signaling molecule in the vascular system: an overview. *J Cardiovasc Pharmacol*, *34*(6), 879-886.

Jahng, J., Jung, I. S., Choi, E. J., Conklin, J. L. & Park, H. (2012). The effects of methane and hydrogen gases produced by enteric bacteria on ileal motility and colonic transit time. *Neurogastroenterol Motil*, *24*(2), 185-e192.

Keppler, F., Boros, M., Frankenberg, C., Lelieveld, J., McLeod, A., Pirttilä, A. M., Röckmann, T. & Schnitzler, J. (2009). Methane formation in aerobic environments. *Environmental Chemistry*, *6*(6), 459-465.

Keppler, F., Hamilton, J. T., Brass, M. & Rockmann, T. (2006). Methane emissions from terrestrial plants under aerobic conditions. *Nature*, *439*(7073), 187-191.

Keppler, F., Hamilton, J. T., McRoberts, W. C., Vigano, I., Brass, M. & Rockmann, T. (2008). Methoxyl groups of plant pectin as a precursor of atmospheric methane: evidence from deuterium labelling studies. *New Phytol*, *178*(4), 808-814.

Koeck, T., Fu, X., Hazen, S. L., Crabb, J. W., Stuehr, D. J. & Aulak, K. S. (2004). Rapid and selective oxygen-regulated protein tyrosine denitration and nitration in mitochondria. *J Biol Chem*, *279*(26), 27257-27262.

Kovacs, T., Varga, G., Erces, D., Tokes, T., Tiszlavicz, L., Ghyczy, M., Boros, M. & Kaszaki, J. (2012). Dietary phosphatidylcholine supplemen-tation attenuates inflammatory mucosal damage in a rat model of experimental colitis. *Shock*, *38*(2), 177-185.

Kremer, M. L. (2003). The Fenton Reaction. Dependence of the Rate on pH. *The Journal of Physical Chemistry A*, *107*(11), 1734-1741.

Lamon, B. D., Zhang, F. F., Puri, N., Brodsky, S. V., Goligorsky, M. S. & Nasjletti, A. (2009). Dual pathways of carbon monoxide-mediated vasoregulation: modulation by redox mechanisms. *Circ Res*, *105*(8), 775-783.

Lee, K. M., Paik, C. N., Chung, W. C., Yang, J. M. & Choi, M. G. (2013). Breath methane positivity is more common and higher in patients with objectively proven delayed transit constipation. *Eur J Gastroenterol Hepatol*, *25*(6), 726-732.

Levitt, M. D., Furne, J. K., Kuskowski, M. & Ruddy, J. (2006). Stability of human methanogenic flora over 35 years and a review of insights obtained from breath methane measurements. *Clin Gastroenterol Hepatol*, *4*(2), 123-129.

Ligor, T., Ligor, M., Amann, A., Ager, C., Bachler, M., Dzien, A. & Buszewski, B. (2008). The analysis of healthy volunteers' exhaled breath by the use of solid-phase microextraction and GC-MS. *J Breath Res*, *2*(4), 046006.

Lowe, D. C. (2006). Global change: a green source of surprise. *Nature*, *439*(7073), 148-149.

McLeod, A. R., Fry, S. C., Loake, G. J., Messenger, D. J., Reay, D. S., Smith, K. A. & Yun, B. W. (2008). Ultraviolet radiation drives methane emissions from terrestrial plant pectins. *New Phytol*, *180*(1), 124-132.

Messenger, D. J., McLeod, A. R. & Fry, S. C. (2009). The role of ultraviolet radiation, photosensitizers, reactive oxygen species and ester groups in mechanisms of methane formation from pectin. *Plant, Cell & Environment*, *32*(1), 1-9.

Mitchell, L., Brook, E., Lee, J. E., Buizert, C. & Sowers, T. (2013). Constraints on the late holocene anthropogenic contribution to the atmospheric methane budget. *Science*, *342*(6161), 964-966.

Moncada, S. & Erusalimsky, J. D. (2002). Does nitric oxide modulate mitochondrial energy generation and apoptosis? *Nat Rev Mol Cell Biol*, *3*(3), 214-220.

Morris, B. E., Herbst, F. A., Bastida, F., Seifert, J., von Bergen, M., Richnow, H. H. & Suflita, J. M. (2012). Microbial interactions during residual oil and n-fatty acid metabolism by a methanogenic consortium. *Environ Microbiol Rep*, *4*(3), 297-306.

Motterlini, R. & Otterbein, L. E. (2010). The therapeutic potential of carbon monoxide. *Nat Rev Drug Discov*, *9*(9), 728-743.

Mukherjee, R. & Chakrabarti, O. (2013). Mitochondrial quality control: decommissioning power plants in neurodegenerative diseases. *Scientific WorldJournal*, 2013, 180759.

Musatov, A. & Robinson, N. C. (2012). Susceptibility of mitochondrial electron-transport complexes to oxidative damage. Focus on cytochrome c oxidase. *Free Radic Res*, *46*(11), 1313-1326.

Ngo, J. K., Pomatto, L. C. & Davies, K. J. (2013). Upregulation of the mitochondrial Lon Protease allows adaptation to acute oxidative stress but dysregulation is associated with chronic stress, disease, and aging. *Redox Biol*, *1*(1), 258-264.

Nohl, H., Breuninger, V. & Hegner, D. (1978). Influence of Mitochondrial Radical Formation on Energy-Linked Respiration. *European Journal of Biochemistry*, *90*(2), 385-390.

Nose, K., Nunome, Y., Kondo, T., Araki, S. & Tsuda, T. (2005). Identification of gas emanated from human skin: methane, ethylene, and ethane. *Anal Sci*, *21*(6), 625-628.

Ohsawa, I., Ishikawa, M., Takahashi, K., Watanabe, M., Nishimaki, K., Yamagata, K., Katsura, K., Katayama, Y., Asoh, S. & Ohta, S. (2007). Hydrogen acts as a therapeutic antioxidant by selectively reducing cytotoxic oxygen radicals. *Nat Med*, *13*(6), 688-694.

Phelan, A. M. & Lange, D. G. (1991). Ischemia/reperfusion-induced changes in membrane fluidity characteristics of brain capillary endothelial cells and its prevention by liposomal-incorporated superoxide dismutase. Biochimica et Biophysica Acta (BBA) – *Biomembranes*, *1067*(1), 97-102.

Pimentel, M., Lin, H. C., Enayati, P., van den Burg, B., Lee, H. R., Chen, J. H., Park, S., Kong, Y. & Conklin, J. (2006). Methane, a gas produced by enteric bacteria, slows intestinal transit and augments small intestinal contractile activity. *Am J Physiol Gastrointest Liver Physiol*, *290*(6), G1089-1095.

Rochette, L., Cottin, Y., Zeller, M. & Vergely, C. (2013). Carbon monoxide: mechanisms of action and potential clinical implications. *Pharmacol Ther*, *137*(2), 133-152.

Russo, I., Del Mese, P., Viretto, M., Doronzo, G., Mattiello, L., Trovati, M. & Anfossi, G. (2008). Sodium azide, a bacteriostatic preservative contained in commercially available laboratory reagents, influences the responses of human platelets via the cGMP/PKG/VASP pathway. *Clin Biochem*, *41*(4-5), 343-349.

Ryter, S. W. & Otterbein, L. E. (2004). Carbon monoxide in biology and medicine. *Bioessays*, *26*(3), 270-280.

Sarti, P., Arese, M., Bacchi, A., Barone, M. C., Forte, E., Mastronicola, D., Brunori, M. & Giuffre, A. (2003). Nitric oxide and mitochondrial complex IV. *IUBMB Life*, *55*(10-11), 605-611.

Serra, M. C. C., Pessoa, F. L. P. & Palavra, A. M. F. (2006). Solubility of methane in water and in a medium for the cultivation of methanotrophs bacteria. *The Journal of Chemical Thermodynamics*, *38*(12), 1629-1633.

Sheu, S.-S, Nauduri, D. & Anders, M. W. (2006). Targeting antioxidants to mitochondria, A new therapeutic direction. *Biochimica et Biophysica Acta (BBA) - Molecular Basis of Disease*, *1762*(2), 256-265.

Strocchi, A., Furne, J., Ellis, C. & Levitt, M. D. (1994). Methanogens outcompete sulphate reducing bacteria for H2 in the human colon. *Gut*, *35*(8), 1098-1101.

Szabo, C., Coletta, C., Chao, C., Modis, K., Szczesny, B., Papapetropoulos, A. & Hellmich, M. R. (2013). Tumor-derived hydrogen sulfide, produced by cystathionine-beta-synthase, stimulates bioenergetics, cell proliferation, and angiogenesis in colon cancer. *Proc Natl Acad Sci U S A*, *110*(30), 12474-12479.

Tappia, P. S., Dent, M. R. & Dhalla, N. S. (2006). Oxidative stress and redox regulation of phospholipase D in myocardial disease. *Free Radic Biol Med*, *41*(3), 349-361.

Triantafyllou, K., Chang, C. & Pimentel, M. (2014). Methanogens, Methane and Gastrointestinal Motility. *J Neurogastroenterol Motil*, *20*(1), 31-40.

Tuboly, E., Szabo, A., Eros, G., Mohacsi, A., Szabo, G., Tengolics, R., Rakhely, G. & Boros, M. (2013a). Determination of endogenous methane formation by photoacoustic spectroscopy. *J Breath Res*, *7*(4), 046004.

Tuboly, E., Szabo, A., Garab, D., Bartha, G., Janovszky, A., Eros, G., Szabo, A., Mohacsi, A., Szabo, G., Kaszaki, J., Ghyczy, M. & Boros, M. (2013b). Methane biogenesis during sodium azide-induced chemical hypoxia in rats. *Am J Physiol Cell Physiol*, *304*(2), C207-214.

VanUffelen, B. E., Van der Zee, J., de Koster, B. M., VanSteveninck, J. & Elferink, J. G. (1998). Sodium azide enhances neutrophil migration and exocytosis: involvement of nitric oxide, cyclic GMP and calcium. *Life Sci*, *63*(8), 645-657.

Vigano, I., Röckmann, T., Holzinger, R., van Dijk, A., Keppler, F., Greule, M., Brand, W. A., Geilmann, H. & van Weelden, H. (2009). The stable isotope signature of methane emitted from plant material under UV irradiation. *Atmospheric Environment*, *43*(35), 5637-5646.

Vigano, I., van Weelden, H., Holzinger, R., Keppler, F., McLeod, A. & Röckmann, T. (2008). Effect of UV radiation and temperature on the emission of methane from plant biomass and structural components. *Biogeosciences*, *5*(3), 937-947.

Wang, B., Hou, L., Liu, W. & Wang, Z. (2013a). Non-microbial methane emissions from soils. *Atmospheric Environment*, *80*, 290-298.

Wang, J., Lu, S., Moenne-Loccoz, P. & Ortiz de Montellano, PR. (2003). Interaction of nitric oxide with human heme oxygenase-1. *J Biol Chem*, *278*(4), 2341-2347.

Wang, Z-P, Chang, S. X., Chen, H. & Han, X-G. (2013b). Widespread non-microbial methane production by organic compounds and the impact of environmental stresses. *Earth-Science Reviews*, *127*(0), 193-202.

Wang, Z., Keppler, F., Greule, M. & Hamilton, J. T. G. (2011). Non-microbial methane emissions from fresh leaves: Effects of physical wounding and anoxia. *Atmospheric Environment*, *45*(28), 4915-4921.

Wishkerman, A., Greiner, S., Ghyczy, M., Boros, M., Rausch, T., Lenhart, K. & Keppler, F. (2011). Enhanced formation of methane in plant cell cultures by inhibition of cytochrome c oxidase. *Plant Cell Environ*, *34*(3), 457-464.

Wu, L. & Wang, R. (2005). Carbon monoxide: endogenous production, physiological functions, and pharmacological applications. *Pharmacol Rev*, *57*(4), 585-630.

Yu, Y. & Pawliszyn, J. (2004). On-line monitoring of breath by membrane extraction with sorbent interface coupled with CO2 sensor. *J Chromatogr A*, *1056*(1-2), 35-41.

Zakhari, S. (2013). Alcohol metabolism and epigenetics changes. *Alcohol Res*, *35*(1), 6-16.

In: Methanogenesis
Editor: Gagik Badalians Gholikandi

ISBN: 978-1-63321-567-2

Chapter 3

METHANE PRODUCTION YIELD AND PERFORMANCE ASSESSMENT OF CONVENTIONAL AND ELECTROLYSIS-ENHANCED ABR (EABR)

***Gagik Badalians Gholikandi*[1*] *and Shervin Jamshidi*[2]**

[1]Faculty of Water Engineering and Environment,
Shahid Beheshti University, A.C., Tehran, Iran
[2]Water and Wastewater Research Centre (WWRC),
Water Research Institute (WRI), Tehran, Iran

ABSTRACT

This study is aimed to examine the anaerobic baffled reactor (ABR) performance and find an engineered enhancement, with emphases on methane yield in decisive operational conditions. For this purpose, at first, anaerobic wastewater treatment has been defined as a dynamic three steps process having fermentation, equalization, and methane production in order. Also, particular operational indices were introduced to track the symptoms of these steps. Then, two pilots with 6 compartments, having net volume of 60 litres, were used and inoculated with dairy wastewater. Based on experimental results, it was observed that in 36 and 18 hours

* Corresponding author: Hakimieh- Tehranpars, Shahid Abbaspour Blvd., PO Code 16589-53571, PO Box 16765-1719, Tehran, Iran. Email: g.badalians@yahoo.com.

retention time, COD reduction was about 82% and 77% while the methane composition of the biogas was 61% and 53%, respectively. It shows that the hydraulic loads can adequately be controlled in conventional ABR with proper configuration without imposing considerable threat for methane production. In contrary, it was observed that instantaneous reduction in pH value of influent to lower than 6 significantly reduced the total COD removal efficiency, the biogas production yield and methane composition. Consequently, the conventional ABR was recommended to be modified by electrolysis application (EABR). An intermittent usage of this enhancement could increase total COD removal efficiency to 69% with 48% methane composition in the biogas. As a conclusion, it can be claimed that an engineered ABR would efficiently and steadily be used for organic removal or biogas generation due to partial separation of three steps of anaerobic process even in lower HRTs. In cases of low pH, the electrolysis enhanced ABR (EABR) was verified to be applicable to maintain the activity of methanogenesis bacteria from being disturbed and to assist in equalization before methane production.

Keywords: Anaerobic baffled reactor, Biogas, Methane, Electrolysis, pH, hydraulic retention time

3.1. INTRODUCTION

However many attractive features can be offered by anaerobic systems, i.e., low operating and maintenance costs, biogas production, low energy and labour requirement, low sludge production, and high efficiency of organics removal; they unfortunately encountered unwarranted reputation as an unstable and difficult to control process (Gerardi, 2003; Leitao, et al. 2006). This can mainly be due to several reasons including lack of adequate knowledge of anaerobic process and microbiology, unavailability of operational data, susceptibility and complexity of the process, and few operator-friendly approaches controlling, monitoring or even upgrading reactor efficiency. Since operational stability is as essential as organic removal efficiency, an engineered environment should be introduced to reduce the disturbing factors of operational upsets and their ramification. For this purpose, at first it was recommended that anaerobic process, effective operational conditions, and indicators should be analyzed and reviewed. This assists to discuss in detail about the advantages or disadvantages of suggested engineered solutions. Besides, this will outline the bottlenecks of operational upsets on anaerobic digestion and methane formation.

3.1.1. Process Analysis

In anaerobic process, organic compounds including carbohydrates, proteins or lipids in wastewater, are converted to biogas in a series of bacterial events. In literatures, these events are commonly defined in 2 to 4 stages containing different activities (Hutnan, et al. 1999; Lyberatos and Skiadas, 1999). This study is focused on sequencing three process-operational stages having influence on the performance of anaerobic systems in wastewater treatment. These are termed here as fermentation, equalization, and methane production in order described in details with emphasizes on operational scope.

1) ***Fermentation*** - Once organic compounds are inoculated to the anaerobic systems, the biopolymers are gradually hydrolyzed and fermented to volatile fatty acids (VFAs), hydrogen and carbon dioxide by various acid-forming bacteria found in the mixed liquor (Nielsen, et al. 2007). There, some facultative bacteria ferment simple and soluble organics while the others require strictly obligate anaerobic condition for complex proteins and carbohydrates fermentations. Therefore, the primary by products would totally be relied on either the type of bacteria involved in the fermentation process or substrate itself (Gerardi, 2003). This is also dependent on variations of operational conditions including hydraulic retention time and the influent characteristics (i.e., pH, oxidation-reduction potential or chemical oxidation demand). In low hydraulic retention times, or in slowly degrading influents such as pulp and paper industries, or even in toxics and heavy metal existence, and low temperature, often this step is introduced as the rate limiting step of the whole process because the probability of suitable contact between substrate, appropriate enzymatic activity and micorganisms with new cell production would be decreased (Colussi, et al. 2009; Merkel, et al. 1999; Palatsi, e al. 2009; Yu and Fang, 2003). Accordingly, it can be summarized that the first step intends to provide more simple organic compounds, such as volatile acids and alcohols, for further steps which is typically resulted in a decrease of pH value to below 7 and oxidation-reduction potential (ORP) to values less than -100 mV

(Gerardi, 2003). This is also proved to be effective on equalization step and hydrogen production as well (Li, et al. 2009). In completely mixed anaerobic systems, or in wastewater having great values of readily biodegradable compounds, such as dairy wastewater, where fermentation is not defined as a rate limiting step, the accumulation of VFAs can inhibit or totally terminate methane production. This can confine the COD removal and VFAs digestion. Therefore, some plug flow or sequencing reactors are introduced to alleviate the adverse effects of VFAs accumulation (Tchobanouglous, et al. 2003).

2) ***Equalization*** - Although, acid forming bacteria produces different compounds of VFAs, acetate is mostly favourable for methane-forming bacteria. Acceptable enzymatic activity of acid-forming bacteria occurs above pH 5, while it is 6.2 for methane-forming bacteria (Gerardi, 2003). Besides, ORP value is required to be decreased to below -300mV for stable reproduction of methane by acetoclastic methanogens. Therefore, an intermission step is required to reduce or neutralize the disturbances and make balance between different microbial activities. Yet, due to accumulation of hydrogen in this step (equation 1), a binary symbiotic relationship between acetate forming bacteria and hydrogenotrophic methanogens takes place to equalize the substrate for acetoclastic methanogens.

$$CH_3CH_2OH + CO_2 \rightarrow CH_3COOH + 2H_2 \quad (1)$$

For this purpose, acetoclastic methane-forming bacteria serve acetate as an essential substrate and their survival are relied on acetate availability, neutral pH and low ORP. However, the bacteria responsible for acetate formation can only survive when hydrogen, as a metabolic product, is continuously removed. Since the partial pressure of hydrogen does not exceed, the activity of bacteria would not be terminated. This is only balanced by hydrogenotrophic methanogens with rather higher kinetic rate of reproduction. Therefore, hydrogen concentrations are recommended to be maintained very low. Accordingly, Voolapalli and stuckey, (2001) have introduced hydrogen and its ratio with formate as a key operational parameter of anaerobic treatment systems. They believed that during shock loads, sampling and analysis of formate and hydrogen can almost instantaneously show that whether the shock load was effective on process or not. Based on experimental results, it was observed that tracing formate in neutral pH (about 7.5) can reveal better operational monitoring while it was recommended to be replaced with hydrogen partial pressure in lower pH values. Besides, Nielsen, et al. (2007) has introduced propionate as a key parameter. They emphasized on this fact that acetate formation (indicating as the ratio of acetate to propionate) can demonstrate that whether process is able to be followed in methanogenesis. Also, the experimental results have shown that propionate were more moderate than any other volatile acids in shock loads while could best be described for process normalization. Since all these circumstances normally occur in the equalization step, it can be concluded that this level would be the challenging operational step with the highest priority in monitoring to maintain the system more stabilized.

3) ***Methane production*** - The only gas of economic value is the methane as a natural, odourless, cleanly burnt and flammable which can be used as a fuel. Methane, as the final result of anaerobic process, is produced by methane-forming bacteria using hydrogen, carbon dioxide and simplistic compounds like methanol, methylamin or acetate (Ferry, 1993; Gerardi, 2003). It should be noticed that the fermentation of propionate and larger fatty acids would result in minor methane production which can be neglected. Thus three major microbial groups of methane-forming bacteria are known as 1) hydrogenotrophic methanogens (equation 2), 2) acetotrophic methanogens (equation 3), and 3) methylotrophic methanogens.

$$CO_2 + 4H_2 \rightarrow CH_4 + 2H_2O \quad (2)$$

$$CH_3COOH \rightarrow CH_4 + CO_2 \quad (3)$$

Since hydrogenotrophic methanogens have the most energy gained by the bacteria among these three microbial groups, they thermodynamically commence on methane production through hydrogen utilization. This maintains the hydrogen partial pressure low required for acetogenic bacteria. However, due to the limited hydrogen supply, it cannot be expected that more than 30% of ultimate potential of methane production would be achieved through this approach. The majority of methane obtained from acetate mostly produced by two genera of acetotrophic methanogens, *Methanosarcina* and *Methanotrix* (Gerardi, 2003; Krishna, et al.

2009). Accordingly, in the best operational performance of anaerobic systems, it is anticipated that carbon dioxide is utilized in the equalization step by acetate forming bacteria and hydrogenotrophic methanogens. As a results, the ratio of methane to carbon dioxide increases in the biogas. In literatures, it was estimated that approximately 65% and 35% of biogas volume are attributed to methane and carbon dioxide, respectively (Tchobanouglous, et al. 2003). Therefore, the ratio of methane to carbon dioxide can be introduced as an indicator showing the progress of methane formation and equalization of the second step. Surely, poor operation, adverse environmental conditions, and systems imbalance would reduce the volume of biogas or particularly methane ratio. Therefore, the process analysis implies that anaerobic reactors efficiency, stability and performance are relied on well symbiotic relationship between bacteria in the second step. In that order, their interrelationships play an important role on either stable or imbalanced performance of anaerobic systems. Thus, the monitoring and upgrading solution of operation should consider the second step thoroughly.

It should be noted that predicting what happens within the anaerobic process and the formation of biogas or microbial responses to different operational conditions is a challenge which was previously simulated by different researches. For instance, hierarchical network was used for monitoring anaerobic process interrelations (Holubar, et al. 2002), while the microbial responses to pH variations were simulated through artificial neural network as well (Horiuchi, et al. 2001). Besides, methane formation tracing in the reactor was modelled by artificial neural network (Shi, et al. 2009; Strik, et al. 2005) and fuzzy logic based systems (Turkdogan-Aydinol and Yetilmezsoy, 2010). Furthermore, some were focused on using binary simulation techniques of artificial neural network and genetic algorithm to predict hydrogen production (Mu and Yu, 2007; Yang and Han-Qing, 2007).

Since anaerobic process takes place as a result of the combination of different microbial activities, it is recommended to be analyzed or even simulated dynamically. Therefore, in Figure 1, the relations of fundamental microbial activities in municipal wastewater treatment or sewage sludge digestion have been demonstrated using causal loops of system dynamics approach with Vensim PLE software. However, the purpose of this study is not to simulate or predict chemical reactions, it can be illustrated in simple causal loops to describe the relationships graphically (Fabien, 2009). Accordingly, the whole anaerobic process analysis and the microbial interactions mentioned before in the manuscript is summarized in Figure 1. It should be noted that markers with plus sign shows direct relation between the material and product. In other words, the more acetate concentration, the more methane gas produced (B4). But, this process can reduce acetate concentration balancing the reactions (the minus signed marker). Likewise, the more hydrogen and carbon dioxide, the more possibility of methanogenesis. Furthermore, the processes of B3 and B4 can directly lead into more development in methane production while B1 and B2 must be monitored and controlled to avoid accumulating of by-products such as hydrogen and VFAs.

3.1.2. Operational Indicators

To monitor and find out which step is responsible inhibiting process accomplishment, several indicators of operational upsets were developed. These are either increases or decreases in specific conditions. For example, as mentioned before, biogas and particularly methane production in association with methane to carbon dioxide ratio focuses on answering this question that whether the second or third step is commenced or accomplished. Likewise, the variations in pH values or volatile fatty acid concentrations can focus on interpreting that whether the first or second step is followed. Meanwhile, since the enzymatic energy is mostly obtained by oxidation-reduction reactions, the spectrum of ORP values can be introduced as an index to find out which bacteria group or chemical reaction would be possibly active and dominated in the mixed liquor. Since the rate limiting phase of anaerobic process is usually attributed to the conversion of volatile fatty acids to methane, the activity of methane-forming bacteria, and particularly the performance on equalization step can be determined by measurement of these indicators (Li, et al. 2009). However, VFAs concentrations and methane formation were typically determined as the most crucial indicators, It should be noted that the former is meticulously highlight the current condition of the mixed liquor (except propionate) and is totally sensitive to environmental and operational variation, while the latter illustrates the overall performance of the system in longer periods and is more correlated to the chemical oxidation demand (COD) removal. Therefore, it is recommended that the second indicator is used for monitoring operational stability of engineered systems due to its correlation to anaerobic efficiency. For example, most of organic compounds of influent is used for metabolic reactions and released as biogas production. Due to low sludge production, most of carbonaceous biochemical oxidation demands (CBOD) would be in form of biogas particularly CO_2 and CH_4. Hence, it is expected that microbial activities of the first and second steps would totally not able to exceed 50% in COD removal (Lyberatos and Skiadas, 1999; Merkel, 1999; Tchobanoglous, et al. 2003). Therefore, determination of COD removal rate, biogas production rate, the ratio of methane to carbon dioxide and ORP can together indicate either the process efficiency or operational reliability of anaerobic systems generally in long terms and were recommended to be monitored periodically and assessed for optimum performance.

3.1.3. Operational Conditions

Different conditions are directly or indirectly in charge of unstable function of anaerobic systems. The two typical factors are hydraulic overload and pH variations. Generally, pH affects on the enzymatic activity of some microbial groups, but of all the operational conditions within an anaerobic system, for example, temperature, solids concentration, and volatile solids of the feed sludge; HRT is perhaps the most important operational condition affecting the conversion of volatile solids to gaseous product (Gerardi, 2003). This ensures the generation time required for a population of bacteria to double in size as well as controlling solids wash-out. In contrary, the metabolic activities of different bacteria and rapid hydrolysis require that they be in close spatial contact. This can be provided by mixing or in rather low HRTs. In order to reduce hydraulic retention times of anaerobic systems and increase its organic removal efficiency and solids retention time (SRT), high rate systems such as upflow anaerobic sludge bed (UASB) and anaerobic baffled reactor (ABR) have been introduced as a promising solution for a variety of wastewater treatments. These can result in maximizing removal and buffering capacity for protection against the effects of shock loads and toxic compounds as well. For example, in a recent study, an intermittent operation of UASB was recommended to be followed for stable biogas production in the system (Nadais, et al. 2011).

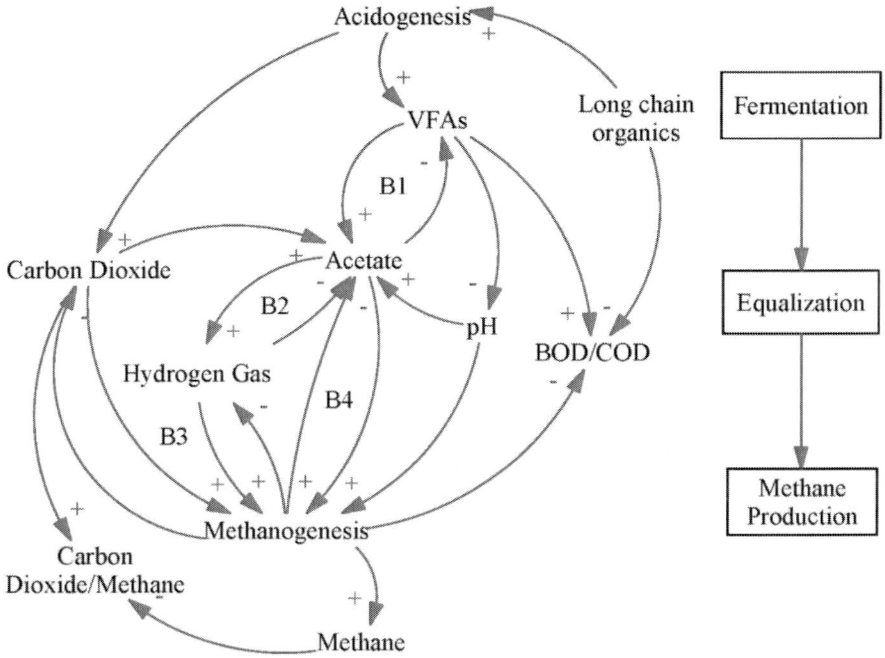

Figure 1. Causal loops of primary parameters in anaerobic process.

3.1.4. ABR

Stable and practical operation of biological units in wastewater treatment is relied on experimental results of pilot scale studies carried out in advance. For example, to upgrade the configuration of conventional systems, they should be verified in different operational conditions to find a most compatible design. Due to the susceptibility of conventional anaerobic reactors, typically high rate systems, such as ABR, are selected for experimentations (Badalians Gholikandi, et al. 2014). Conventional anaerobic baffled reactor (ABR) was identified as a promising solution for urban wastewater treatment. This system is typically defined as a hydraulically based reactor forcing wastewater to be in direct contact with the biomass (Wang, et al. 2004; Sarathai, et al. 2010). The primary advantage of ABR is partial separation of microbial groups longitudinally down the reactor and more stable operation (Nachaiyasit and Stuckey, 1995; Badalians Gholikandi, et al. 2011). This can facilitate anaerobic process analysis using typical operational factors and indicators. In Figure 2, the schematic separation of microbial consortium and sequencing process in ABR is illustrated. Based on system capability in optimized configuration, it can be assumed that fermentation, equalization, and methane production take effect in initial, middle, and final compartments, respectively. There, it is expected that operational quality parameters indicate and verify the status of anaerobic process. As shown in Figure 2, the quality parameters mentioned on top would be increased in each step while those depicted on the bottom would be reduced. Therefore, they can be used as an index for process analysis. For example, in initial compartment, the long chain organics (i.e., proteins or carbohydrates) are hydrolyzed where acidogenesis bacteria produce volatile fatty acids. Thus, the value of VFA ratio to alkalinity would be increased while pH is reduced. This hydrolysis can even increase the proportion of BOD to COD in the mixed liquor by discharging VFAs and soluble microbial products (SMPs). In middle compartments, the values of by-products, such as hydrogen and acetate, are increased where biogas is mostly made of carbon dioxide. In these two steps, ORP is continuously reduced to provide obligate anaerobic condition for the third step. In final compartments, it probably can be verified that by digestion of VFAs, acetate, carbon dioxide, and hydrogen by different methanogens, would decrease the values of by-products where methane is increased significantly in biogas.

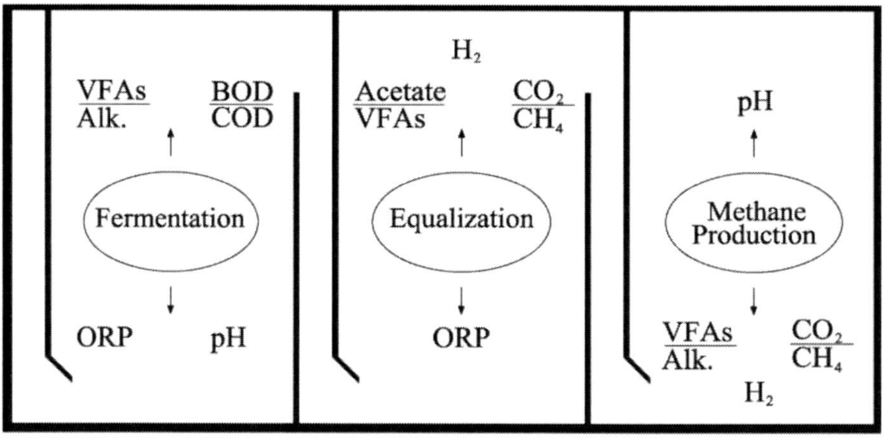

Figure 2. The schematic separation of anaerobic process steps within ABR showing objective operational indices monitoring each step. The upward and downward marker respectively indicates an increase or decrease in parameters.

Another explicit specifications mentioned in literatures are as its potential of retaining proper biomass within the system by means of hydraulic based function, and the ability of influent equalization providing microbial evolution (Barber and Stuckey, 1999; Badalians Gholikandi, et al. 2014).

Previously, various researches have been assigned to analyze the performance of ABR treating different types of wastewater. For example, low strength municipal wastewater has been extensively studied, and it was demonstrated that ABR has the potential of being practiced efficiently (Angenent, et al. 2001; Bodkhe, 2009; Foxon, et al. 2004; Krishna and Kumar, 2009; Langenhoff and Stuckey, 2000; Sabry, 2010). Regarding to the research results, it was concluded that ABR can be used in this field efficiently either as an independent or preliminary unit. The latter can be accomplished by stabilization ponds (Jamshidi and Badalians Gholikandi, 2014), subsurface horizontal and vertical constructed wetlands (Singh, et al. 2009), duckweed ponds (Nasr, et al. 2009), activated sludge (Kassab, et al. 2010) or membrane systems (Pillay, et al. 2008). Also, alcohol-containing desizing wastewater of textile industries (Liu, et al. 2010a), heavy oil and salty wastewater (Ji, et al. 2009), swine wastewater (Pereira, et al. 2010), high sulfate wastewater (Saritpongteeraka and Chaiprapat, 2008), acidic zinc-containing wastewater (Bayrakdar, et al. 2009), Soybean protein processing wastewater (Zhu, et al. 2008), dilute aircraft de-icing fluid (Marin, et al. 2010), and dilute soluble and colloidal wastewater (Langenhoff and Stuckey, 2000) were investigated thoroughly. Meanwhile, treating dairy wastewater was studied by Tawfik, et al. (2008) on UASB reactor while assessment of ABR performance treating this type of wastewater can be known as a novel approach. Besides, upgrading and modification of ABR (Liu, et al. 2010b) by means of hybrid composition using bamboo carrier (Feng, et al. 2008), periodical (Liu, et al. 2007; Skiadas, et al. 2000) and split feed inoculation (Sallis and Uyanik, 2003) were studied as well. Moreover, the impacts of polymer addition (Uyanik, et al. 2002), mesophilic temperature condition (Ghaniyari-Benis, et al. 2010), start-up period (Hutnan, et al. 1999) hydraulic aspects and operational challenges on its performance and microbial specifications were also evaluated (Badalians Gholikandi, et al. 2011; Ji, et al. 2009; Langenhoff, et al. 2000; Zhu, et al. 2008). Accordingly, it can be summarized that the analysis and evaluation of the effects of different conditions in biologically based systesm, particularly anaerobic baffled reactor was totally set on laboratory and pilot studies in advance. In this case, upgrading, modifications, and finding the most appropriate design and operation can be viable.

3.1.5. Electrolysis

Recently, laboratory scale experiments were carried out to find a substitution for operational control and chemical addition for pH recovery in anaerobic systems. Based on electrolysis specifications, it can be assumed that the performance of electrolytic cell would be relied on electrolyte characteristics and conductivity, electrodes material, the space between the electrodes, electrical potential, and the period of electrolysis (Monk, 2001; Wang, 2006).

As the wastewater is electrolyzed, the ionization of water molecule would take effect where hydrogen ions adsorbed to the cathode and hydroxides would come close to the anode. Accordingly, a pH gradient is formed inside the cell. For example, in typical domestic wastewater, with 12 volts, in approximately less than 2 hours, the pH was observed to be decreased to below 2 in cathode vicinity while anode can develop the pH to more than 12. This is shown in Equation 4. In cathode, hydrogen ions would instantaneously be neutralized and hydrogen gas discharged (equation 5) while carbonates can be reduced to carbon dioxide in anode (equation 6). Based on experimental results, it was observed that pH of the solution and alkalinity values were simultaneously decreased significantly. However, the pH variation was much more than alkalinity changes. Furthermore, carbon dioxide foam was formed on the surface. In electrolysis period, the amount of carbon dioxide and hydrogen gas discharged would be much more than the amount of solubilisation through equation 7. Once the electricity is diffused, the pH value of the solution would be more than conditions before electrolysis. Besides, two essential compounds of equalization step in anaerobic process can be totally varied in solution. This may have a great effect on methanogenesis and rate limiting step of process which theoretically motivates its application to upgrade anaerobic systems (Badalians Gholikandi, et al. 2013).

$$H_2O \rightarrow H^+ + OH^- \qquad (4)$$

$$2H^+ + 2e^- \rightarrow H_2 \qquad (5)$$

$$CO_3^{2-} - 2e^- \rightarrow CO_2 \qquad (6)$$

$$CO_2 + H_2O \leftrightarrow H_2CO_3 \leftrightarrow H^+ + HCO_3^- \qquad (7)$$

Similar technology of electrolysis, as an electrochemical approach, has been used previously for sludge pre-treatment and dewatering. Based on experimental results, it can be concluded that ionization in electrolyte (sewage sludge) can extract the water trapped within the biomass while some biodegradable parts would be dismantled to be digested further in anaerobic sludge digesters (Song, et al. 2010; Yuan, et al. 2011). The latter is totally reported to be relied on electric potential and the volume of electrolyte in association of electrolysis period (Badalians Gholikandi, et al. 2013). Consequently, using electrolysis cell has been testified for enhancement of sludge digestion, particularly as a pre-treatment unit where the ratio of BOD to COD would be increased. Besides, it was proved that this technique has not detrimental effects on biomass and even macrophytes growth (Azizi, et al. 2014).

3.1.6. Conclusion

Based on discussions, it can be concluded that the efficient and stable operation of anaerobic process and reactors, particularly in higher COD influent, are totally relied on the equalization step to not only provide appropriate environmental conditions in pH and ORP, but also to generate acetate and reduce hydrogen partial pressure. All should be provided to enforce acetoclastic methanogenesis be commenced. As a result, the operational indicators of pH and ORP, as the process tracing tools in association with VFA concentrations can focus on the process in details while the ratio of methane to carbon dioxide formation and COD removal rate can outline the overall performance. Since the effects of operational conditions of pH variation and hydraulic retention time in industrial wastewater treatment is totally required to be analyzed in long terms, the VFA concentration is suggested to be removed from operational indicators. For this purpose, the higher values of methane to carbon dioxide can represent a more transition from step 2 to 3.

Regarding to the literatures, in order to compare the biogas production rate of anaerobic systems in different operational conditions, conventional anaerobic baffled reactor treating dairy wastewater were determined to be analyzed in pilot scale. Also, as recommended by Badalians Gholikandi, et al. (2013) and Tartakovski, et al. (2011), electrolytic enhanced anaerobic system would be tested to assess the performance on controlling operational conditions. This approach can be introduced as totally novel study on upgrading conventional ABR and methane formation through the system.

The purpose of this research is to study the effects of hydraulic retention time and influent pH depression, as two typical operational upsets, on COD removal, biogas production rate and eventually methane composition in conventional ABR in pilot scale experiments treating dairy wastewater. Then, the performance of electrolytic enhanced ABR (EABR) would be compared to find out whether this development is capable of equalizing operational condition.

3.2. Materials and Methods

3.2.1. Pilot Setup

In order to follow different operational conditions simultaneously, two bench scale anaerobic baffled reactors (ABR) in parallel were setup and operated continuously from April to August, 2013 in ambient temperature (with an average of 28°C). These pilots were made of Plexiglas in total and net volume of 75 L and 60 L, respectively. These systems consisted of equally six compartments through the standard practice with internal size of 94 cm length, 16 cm width, and 50 cm depth (Badalians Gholikandi, et al. 2014). The compartments were divided by vertical baffles having upflow-to-downflow parts by a ratio of 3:1 as recommended in other studies (Barber and Stuckey, 1999; Liu, et al. 2010b). The sampling and sludge discharge valves were located on the side (with 30 cm distance from the bottom) and beneath, respectively. Besides, the biogas formed in each compartment was directly discharged through the exhaust valves located on the top without any mixture with other compartments. Finally, in pilot B, the first compartment was equipped with electrolytic cell using two electrodes made of graphite (2mm), attached to a direct current electric source having 9 volts and 1000 mA (Figure 3). It should be noted that electrolytic cell was in line intermittently for four hours in a week. Previously, it was observed that 2 to 4 hours would be enough for anaerobic process resurrection (Figure 4). In addition, the electrodes had approximately 10 cm distance located in upflow part while were put 5 cm over the surface with 20 cm length. The electrolyte was also the mixed emulsion of sludge and wastewater. The schematic design of EABR is shown in Figure 3. Also, the laboratory experimental results leading into selection of 9 volts and 4 hours performance time for electrolytic cell are shown in Figure 4.

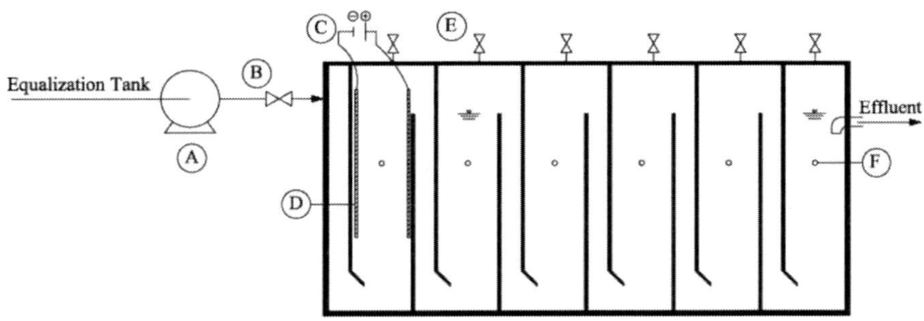

Figure 3. EABR pilot configuration used in this study, A) Peristaltic pump, B) Control valve, C) Electric source, D) Electrode bars, E) Gas exhaust valve, F) Sampling valve.

3.2.2. Pilot Start-up

Start-up was initiated with the batch mode of anaerobic baffled reactors by seeding the mixture of cow dung and sewage sludge at the ratio of 1:2 (50% v/V) on effluent wastewater in each compartment for 2 weeks. There, the content of the reactors was recycled for homogeneity in the same period as per standard practice (Barber and Stuckey, 1999; Hutnan, et al. 1999). Then, the two reactors were operated through continuous inoculation of dairy wastewater starting from 25% loading rate and gradually increased to 100% over 35 days. In this stage, the systems were continued to perform at overall HRT of 52 hours for about 50 days as an acclimatization period. This was performed continuously by peristaltic pumps feeding from the equalization tank of dairy industry wastewater treatment plant. The start-up period was continued till the chemical oxidation demand (COD) removal has reached a constant level while the ratio of the volatile fatty acids to total alkalinity concentration was observed to be maintained below 0.25. Then, the HRTs of both pilots were reduced in two steps and maintained for about one month on 36 hours and then 18 hours. This was carried out to find the effect of HRT on methane formation through conventional ABR. Then, while the system was running in 18 hours retention time, HCl 0.2 N was added to the feeding tanks to reduce the influent pH from the equalization tanks to below 6 (5.8 ±0.2). In this step, pilot B was provided with electrolytic cell and its performance was compared through methane concentration, and COD values in the gas and liquid samples.

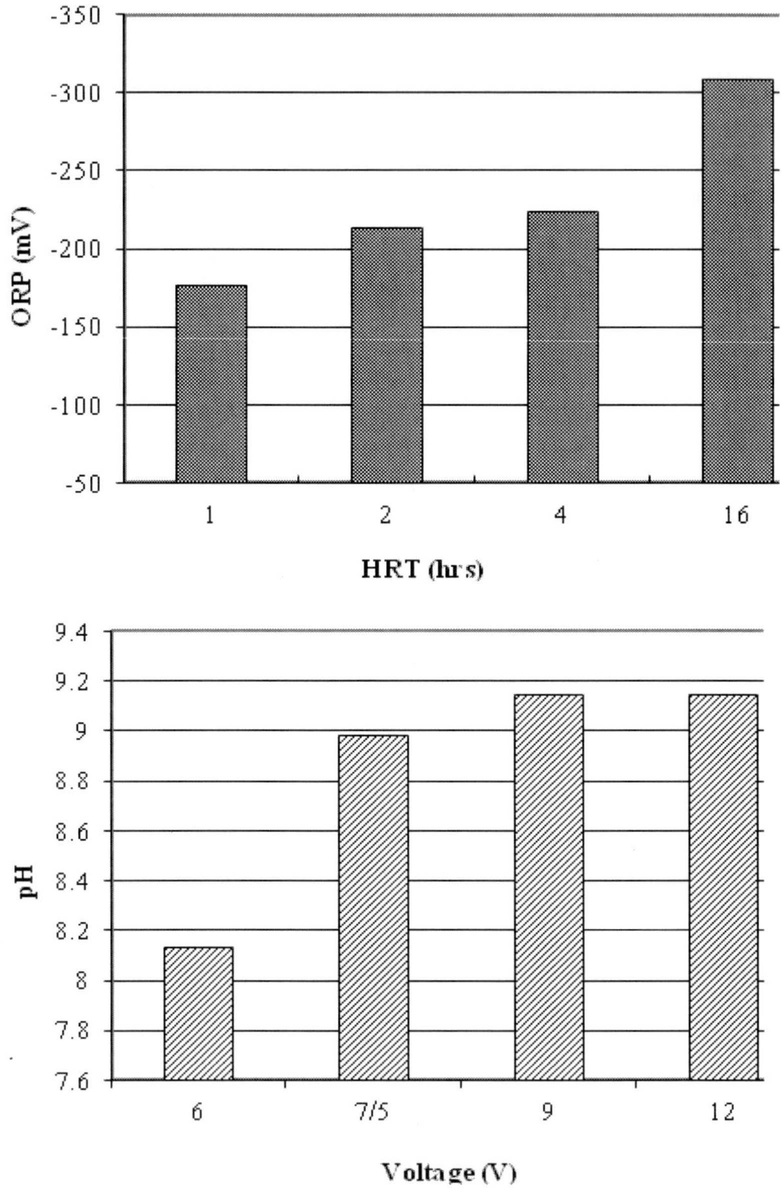

Figure 4. Optimum retention time (left) and voltage (right) for electrolytic cell in laboratory results (Badalians Gholikandi, et al. 2013).

3.2.3. Sampling and Tests

The whole samplings were performed weekly to analyze chemical oxidation demand (COD), pH and ORP values in reactors through all compartments and the influent of the systems. These were monitored based on standard methods (APHA, 2005). The gas composition was also analyzed to find the percentage of CH_4 and CO_2 measured by gas chromatography using thermal conductivity detection (GC-TCD). In order to report the biogas composition, it was assumed that biogas was totally constituted of methane and carbon dioxide. Therefore, other compositions were neglected.

3.3. Results and Discussions

In order to evaluate the effects of operational conditions on the performance of conventional ABR and electrolysis enhanced ABR (EABR), the experimental results were carried out in 2 steps and the results are shown in figures focusing on COD removal rate, pH and ORP values, and biogas composition production rate through the system. As illustrated in Figure 5, it can be concluded that the reduction of HRT from 36 hours retention time to 18 hours could not significantly reduce the overall COD removal of the system. There, it was reduced from the average value of 82% to 77% showing only 5% reductions. However, this difference was about 22% in initial compartment. It shows that the compartmentalized configuration of conventional ABR could be effective against low hydraulic loads and the removal efficiency would be expanded through the system. This was rephrased by artificial neural network (Badalians Gholikandi, et al. 2014). Therefore, it was realized that the 6 compartment ABR was not significantly sensitive to the lower hydraulic rates operation. In contrary, a pH reduction to below 6 (with an average of 5.8) has significantly reduced the overall COD removal of the system. As it is illustrated, pH reduction has lead into a decrease of 48% in overall COD removal (from 77% to 31%). Consequently, pH reduction was verified to have more detrimental effect on ABR performance and its compartmentalized configuration could not compensate its consequences. Therefore, to upgrade ABR to a more robust and flexible system, electrolytic cell was placed in the first chamber to enhance ABR performance. This was termed as electrolytic enhanced ABR (EABR). In this case, it was verified that the electrolytic cell was able to enforce reactor to be maintained approximately in neutral and reliable conditions. Based on Figure 5, it could be realized that the COD removal was instantaneously increased in the following compartments of the electrolytic cell. This recovery was almost completed by compartmentalized configuration of ABR. As a result, the overall COD removal was reported to be only 8% less than typical results of conventional ABR (HRT of 18 hrs). To find the mechanism of EABR performance on reactor recovery, the operational parameters such as ORP and pH were analyzed through the system.

As shown in Figure 6, it is obvious that the average pH value was increased in the first compartment of EABR to a neutral condition. This can be responsible for providing more stabilized microbial conditions in further compartments. Also, it should be considered that electrolysis in the first compartment could significantly reduce the oxidation-reduction potential (ORP) of the system (Figure 7). Providing a neutral pH condition in addition to ORP reduction to a level required for methanogenesis can be introduced as EABR supremacy on a steady operation. This was also proved in batch and laboratory scale experiments as well by Badalians Gholikandi, et al. 2013. To find out the performance of EABR on biogas production, as the final step of the system, the biogas volume was analyzed and mentioned separately as methane and carbon dioxide compositions.

The final product of anaerobic systems is biogas, which was analyzed through the volume produced and its composition. The latter shows that whether methane or carbon dioxide is dominated. The higher the ratio of methane to carbon dioxide, the better and more complete process flows. As demonstrated in Figures 8 to 11, the biogas production rate and methane ratio were differed as a function of operational conditions through the systems.

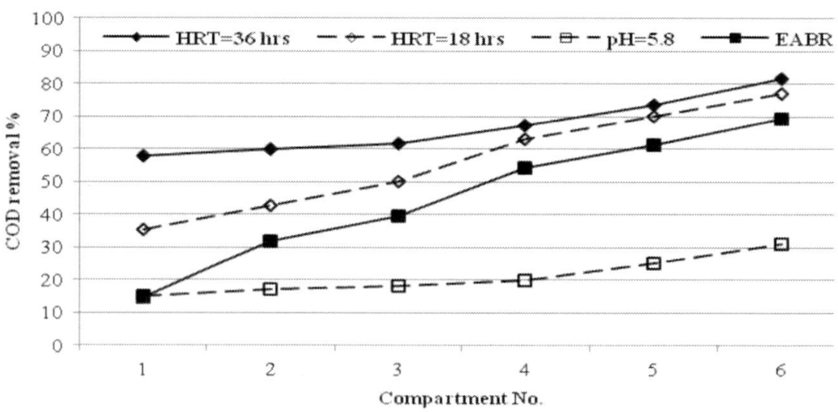

Figure 5. COD removal rate through the systems in different operational conditions.

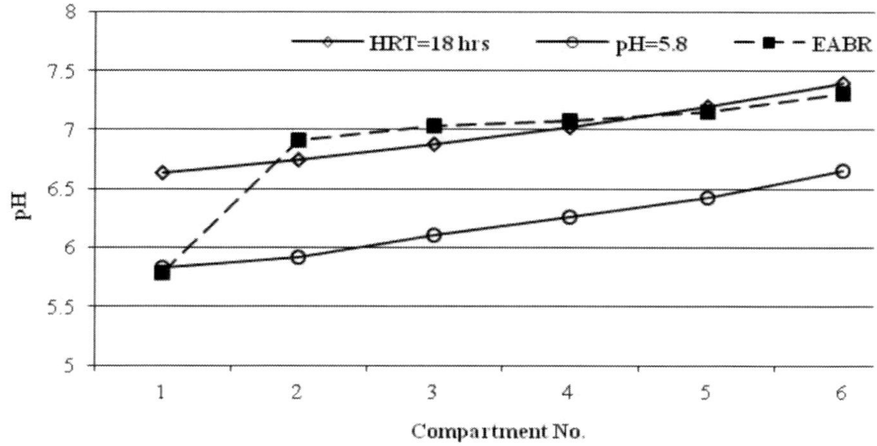

Figure 6. pH variations through the systems in different operational conditions.

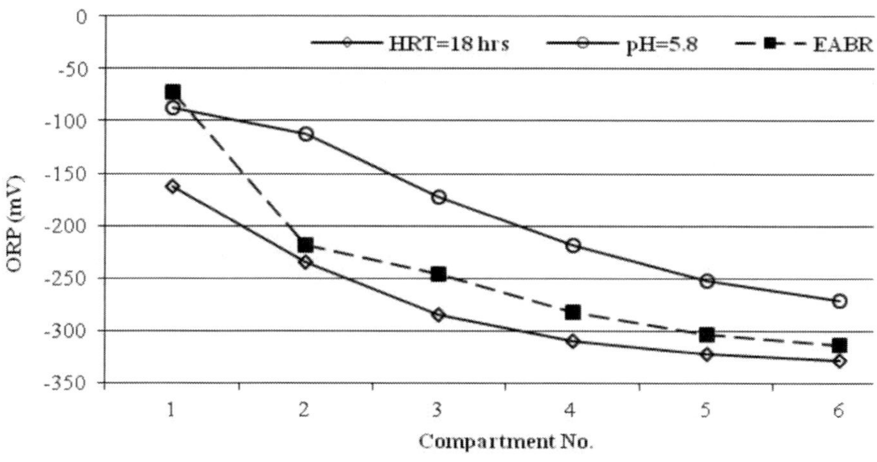

Figure 7. ORP reduction through the systems in different operational conditions.

In 36 hours retention time, due to sufficient time required for methanogens microbial groups within the system, the biogas, even methane, was produced in initial compartments. Moreover, it was observed that methane formation was dominated in compare with carbon dioxide in the second and third compartment which was remained stable to the end (Figure 8). The average biogas production rate was calculated as 0.16 $LL^{-1}d^{-1}$ while methane and carbon dioxide composed of 61% and 39% of biogas, respectively.

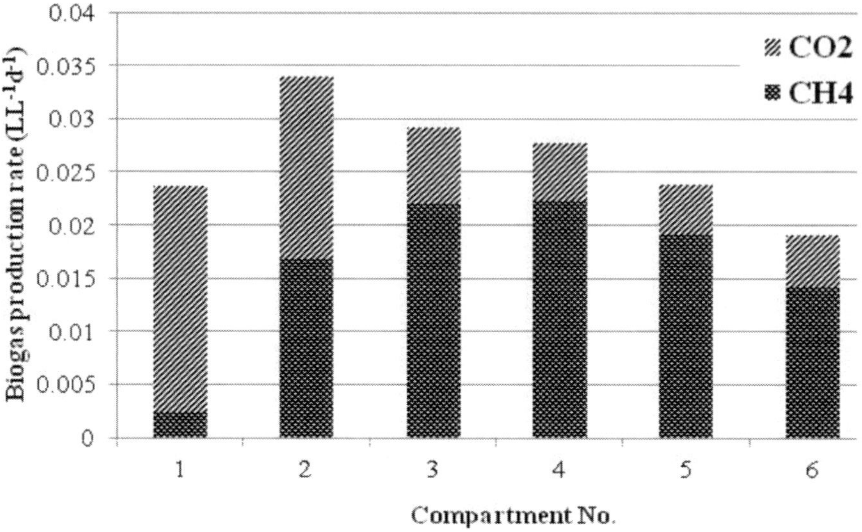

Figure 8. Biogas production rate and its composition within the compartments of ABR with 36 hours retention time.

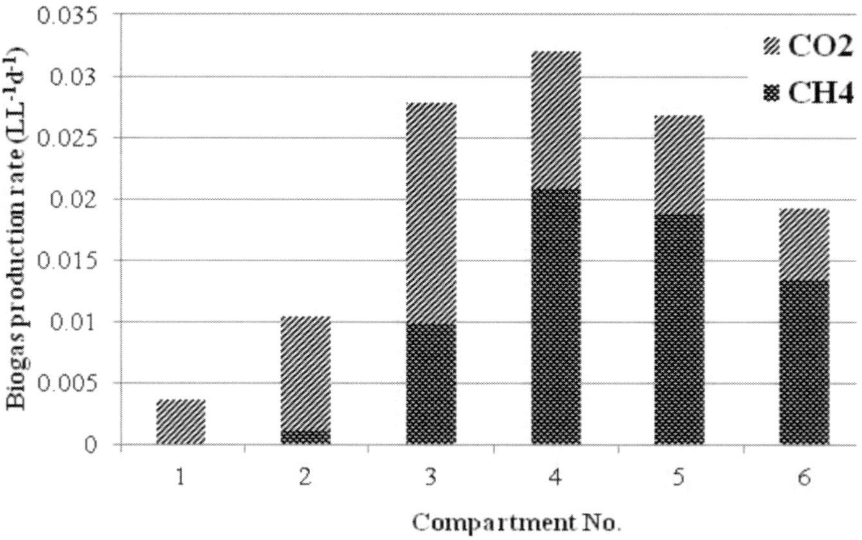

Figure 9. Biogas production rate and its composition within the compartments of ABR with 18 hours retention time.

In contrary, in lower hydraulic retention time of 18 hours, methane production was initiated in the third and dominated in the fourth compartment (Figure 9). It was implied that lowering retention time has resulted in postponing the methane formation and consequently COD removal. This emphasizes on this fact that conventional ABR should necessarily be designed in optimum compartment number, more important than HRT, to not only maintain its COD removal performance in different operational conditions but also make it a more productive system in biogas. In this case, the average biogas production rate was determined as 0.12 $LL^{-1}d^{-1}$ and the methane and carbon dioxide allocation were 53% and 47%, respectively.

The COD removal rate experimentations have revealed that if the pH reduces below 6, this would be a bottleneck for ABR performance. The biogas production analysis has shown the same trend as well. As illustrated in Figure 10, since pH was hardly increased to neutral condition through conventional ABR, biogas production and particularly, methane formation was totally stopped while can only be promoted in final compartments. As calculated, the average biogas production has diminished to 0.02 $LL^{-1}d^{-1}$ which contained less than 15% of methane formation.

Regarding to the values obtained in COD removal of EABR, and with pH and ORP variation trend observed, it can be expected that the biogas production should be much larger than ABR with low influent pH. Regarding to the results of Figure 11, it can be concluded that due to pH development and ORP modifications through electrolytic cell, the methane formation bacteria could successfully reproduce biogas in the third compartment which was dominated in the fifth chamber. The average overall biogas production rate of EABR was estimated about 0.1 $LL^{-1}d^{-1}$ containing 48% of methane in its composition.

In low pH, since the equalization step of anaerobic process cannot kinetically compete with the fermentation step, and consequently it would make an inhibition for methanogenesis, the methane formation would not be dominated. In other word, electrolytic cell through pH and ORP recovery would be able to jump over the equalization step to provide proper environmental conditions for methanogenesis. This can facilitate acetate and methane formation simultaneously. Hence, the COD removal efficiency, methane production and reactor stability would be promoted.

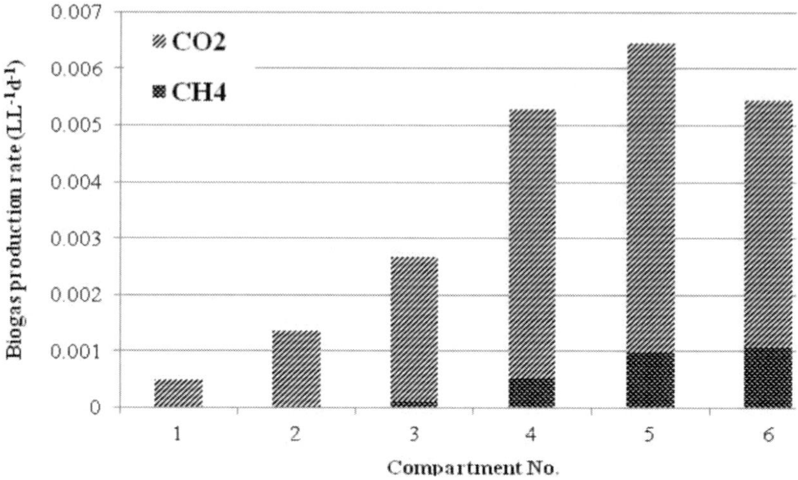

Figure 10. Biogas production rate and its composition within the compartments of ABR with 18 hours retention time and influent pH of 5.8.

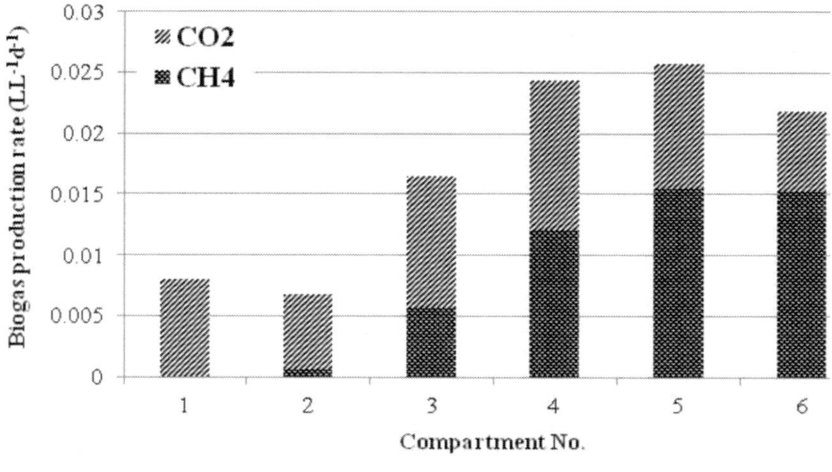

Figure 11. Biogas production rate and its composition within the compartments of EABR with 18 hours retention time and influent pH of 5.8.

CONCLUSION

This study is aimed to focus on the assessment of operational conditions on conventional ABR and develop a solution for unstable function. For this purpose, COD removal and methane formation were determined as two indices for analysis. Regarding to the results, it can be concluded that conventional ABR is much more robust against hydraulic retention time reduction in compare to pH variations. This could be due to compartmentalized configuration of ABR which was not capable of equalizing pH adverse effects on microbial activity as well. In contrary, electrolytic cell was determined as a promising solution for pH recovery, ORP reduction and finally methane formation in the following compartments. Accordingly, electrolytic enhanced ABR (EABR) was introduced as a development upgrading conventional ABR. This was justified through process analysis showing that electrolytic process is capable of providing required environmental conditions for methane forming bacteria without depending on equalization step of acetate forming and hydrogenotroph methane forming bacteria. It obviously can reduce the total kinetic rate of anaerobic process and methane production.

REFERENCES

Angenent, L. T., Banik, G. C. & Sung, S. (2001). Anaerobic migrating blanket reactor treatment of low strength wastewater at low temperature. *Water Environment Research*, *73* (5), 567-574.

APHA. (2005). *Standard Methods for the Examination of Water and Wastewater*. 21tsh ed. American Public Health Association. Washington DC, USA.

Azizi, Sh; Valipour, A., Jamshidi, Sh. & Sithebe, T. (2014). Performance evaluation of the electrolysis process for waste sludge stabilization in decentralization practices. Desalination and Water Treatment, DOI: 10.1080/19443994.2014.886293.

Badalians Gholikandi, G., Jamshidi, S. & Ebrahimzade, V. (2011). The response of anaerobic baffled reactor in various operational conditions. *International journal of Wastewater Treatment and Green Chemistry*, *2*, 1-10.

Badalians Gholikandi, G., Jamshidi, S. & Hazrati, H. (2014). Optimization of anaerobic baffled reactor (ABR) wastewater treatment system using artificial neural network. *Environmental Engineering and Management Journal*, *13* (1), 95-104.

Badalians Gholikandi, G., Jamshidi, S. & Valipour, A. (2013). Application of electrolysis upgrading the operation of anaerobic reactors. *Journal of Environmental Studies*, *38*, 9-16.

Barber, W. P. & Stuckey, D. C. (1999). The use of the anaerobic baffled reactor (ABR) for wastewater treatment: A review. *Water Research*, *33* (7), 1559-1578.

Bayrakdar, A., Sahinkaya, E., Gungor, M., Uyanik, S. & Dilek Atasoy, A. (2009). Performance of sulfidogenic anaerobic baffled reactor (ABR) treating acidic and zinc-containing wastewater. *Bioresource Technology*, *100*, 4354-4360.

Bodkhe, S. Y. (2009). a modified anaerobic baffled reactor for municipal wastewater treatment. *Journal of Environmental Management*, *90*, 2488-2493.

Colussi, I., Cortesi, A., Della Vedova, L., Gallo, V. & Cano Robles, F. K. (2009). Start-up procedures and analysis of heavy metals inhibition on methanogenic activity in EGSB reactor. *Bioresource Technology*, *100*, 6290-6294.

Fabien, B. (2009). *Analytical System Dynamics – Modeling and Simulation*, Springer, New York, USA.

Feng, H., Hu, L., Mahmood, Q., Qiu, C., Fang, C. & Shen, D. (2008). Anaerobic domestic wastewater treatment with bamboo carrier anaerobic baffled reactor. *Journal of International Biodeterioration and Biodegredation*, *62*, 232-238.

Ferry, J. G. (1993). *Methanogenesis: Ecology, Physiology, Biochemistry and Genetics*, Chapman and Hall Microbiology series, Chapman and Hall Inc.

Foxon, K. M., Pillay, S., Lalbahadur, T., Rodda, N., Holder, F. & Buckley, C. A. (2004). the anaerobic baffled reactor (ABR): An appropriate technology for on-site sanitation. *proceeding of Water Institute South Africa (WISA) Biennial Conference*, Cape Town, South Africa.

Gerardi, M. H. (2003). *The microbiology of anaerobic digesters*, A John Wiley & Sons, Inc., Hoboken, New Jersey.

Ghaniyari-Benis, S., Martin, A. & Borja, R. (2010). Kinetic modelling and performance prediction of a hybrid anaerobic baffled reactor treating synthetic wastewater at mesophilic temperature. *Process biochemistry*, *45*(10), 1616-1623.

Holubar, P., Zani, L., Hager, M., Froschl, W., Radak, Z. & Braun, R. (2002). Advanced controlling of anaerobic digestion by means of hierarchical neural networks. *Water Research*, *36*, 2582-2588.

Horiuchi, J. I., Kikuchi, S., Kobayashi, M., Kanno, T. & Shimizu, T. (2001). Modeling of pH response in continuous anaerobic acidogenesis by an artificial neural network. *Biochemical Engineering Journal*, *9*, 199-204.

Hutnan, M., Drtil, M., Mrafkova, L., Derco, J. & Buday, J. (1999). Comparison of startup and anaerobic wastewater treatment in UASB, hybrid and baffled reactor. *Journal of Bioprocess Engineering*, *21*, 439-445.

Jamshidi, Sh. & Badalians Gholikandi, G. (2014). an assessment of using anaerobic baffled reactor to upgrade wastewater stabilization ponds, a pilot study. *International journal of sustainable development and planning*, in press.

Ji, G. D., Sun, T. H., Ni, J. R. & Tong, J. J. (2009). Anaerobic baffled reactor (ABR) for treating heavy oil produced water with high concentrations of salt and poor nutrient. *Bioresource Technology*, *100*, 1108-1114.

Kassab, G., Halalsheh, M., Klapwijk, A., Fayyad, M. & van Lier, J. B. (2010). Sequential anaerobic-aerobic treatment for domestic wastewater – A review. *Bioresource Technology*, *101*, 3299-3310.

Krishna, G. V. T. G., Kumar, P. & Kumar, P. (2009). Treatment of low strength soluble wastewater using an anaerobic baffled reactor (ABR). *Journal of Environmental Management*, *90*, 166-176.

Langenhoff, A., Intrachandra, N. & Stuckey, D. C. (2000). Treatment of dilute soluble and colloidal wastewater using an anaerobic baffled reactor: Influence of hydraulic retention time. *Water Research*, *34* (4), 1307-1317.

Langenhoff, A. & Stuckey, D. C. (2000).Treatment of dilute wastewater using an anaerobic baffled reactor: Effect of low temperature. *Water Research*, *34* (15), 3867-3875.

Leitao, R. C., van Haandel, A. C., Zeeman, G. & Lettinga, G. (2006). The effects of operational and environmental variations on anaerobic wastewater treatment systems: A review. *Bioresource Technology*, *98*, 1105-1118.

Li, J., Zheng, G., He, J., Chang, S. & Qin, Z. (2009). Hydrogen-producing capability of anaerobic activated sludge in three types of fermentations in a continuous stirred-tank reactor. *Biotechnology advances*, *27*, 573-577.

Liu, R. R., Tian, Q., Yang, B. & Chen, J. H. (2010a). Hybrid anaerobic baffled reactor for treatment of desizing wastewater. *International Journal of Environmental Science and Technology*, *7*(1), 111-118.

Liu, R. R., Tian, Q. & Chen, J. (2010b). the developments of anaerobic baffled reactor for wastewater treatment: A review. *African Journal of Biotechnology*, *9*(11), 1535-1542.

Liu, X. L., Ren, N. Q. & Wan, C. L. (2007). Hydrodynamic characteristics of four-compartment periodic anaerobic baffled reactor. *Journal of Environmental Sciences*, *19*, 1159-1165.

Lyberatos, G. & Skiadas, I. V. (1999). Modelling of anaerobic digestion-a review. *Global nest: the international journal*, *1*(2), 63-76.

Marin, J., Kennedy, K. J. & Eskicioglu, C. (2010). Characterization of an anaerobic baffled reactor treating dilute aircraft de-icing fluid and long term effects of operation on granular biomass. *Bioresource Technology*, *101*, 2217-2223.

Merkel, W., Manz, W., Szewzyk, U. & Krauth, K. (1999). Population dynamics in anaerobic wastewater reactors: modelling and in situ characterization. *Water Research*, *33*(10), 2392-2402.

Monk, P. (2001). *Fundamentals of Electroanalytical Chemistry*, Wiley & Sons, Inc. West Sussex, UK.

Mu, Y. & Yu, HQ. (2007). Simulation of biological hydrogen production in a UASB reactor using neural network and genetic algorithm. *International Journal of Hydrogen Energy*, *32*, 3308-3314.

Nadais, H., Barbosa, M., Capela, I., Arroja, L., Ramos, C. G., Grilo, A., Sousa, S. A. & Leitao, J. H. (2011). Enhancing wastewater degradation and biogas production by intermittent operation of UASB reactors. *Energy*, *36*(4), 2164-2168.

Nasr, F. A., Doma, H. S. & Nassar, H. F. (2009). Treatment of domestic wastewater using an anaerobic baffled reactor followed by a duckweed pond for agricultural purposes. *Journal of Environmentalist*, *29*, 270-279.

Nachaiyasit, S. & Stuckey, D. C. (1995). Microbial response to environmental changes in an Anaerobic Baffled Reactor (ABR). *Antonie Van Leeuwenhoek*, *65*, 111-123.

Nielsen, H. B., Uellendahl, H. & Ahring, B. K. (2007). Regulation and optimization of the biogas process: propionate as a key parameter. *Biomass and Bioenergy*, *31*, 820-830.

Palatsi, J., Laureni, M., Andres, M. V., Flotats, X., Nielsen, H. B. & Angelidaki, I. (2009). Strategies for recovering inhibition caused by long chain fatty acids on anaerobic termophilic biogas reactor. *Bioresource Technology*, *100*, 4588-4596.

Pereira, E. L., Montenegro Campos, C. M. & Moterani, F. (2010). Physical-Chemical and operational performance of an anaerobic baffled reactor (ABR) treating swine wastewater. *Acta Scientiarum Technology*, *32*(4), 399-405.

Pillay, S., Foxon, K. M. & Buckley, C. A. (2008). an anaerobic baffled reactor/membrane bioreactor (ABR/MBR) for on-site sanitation in low income areas. *Desalination*, *231*, 91-98.

Sabry, T. (2010). Evaluation of decentralized treatment of sewage employing Upflow Septic Tank/Baffled Reactor (USBR) in developing countries. *Journal of Hazardous Materials*, *174*, 500-505.

Sallis, P. J. & Uyanik, S. (2003). Granule development in a split-feed anaerobic baffled reactor. *Bioresource Technology*, *89*, 255-265.

Sarathai, Y., Koottatep, T. & Morel, A. (2010). Hydraulic characteristics of an anaerobic baffled reactor as onsite wastewater treatment system. *Journal of Environmental Sciences*, *22*(9), 1319-1326.

Saritpongteeraka, K. & Chaiprapat, S. (2008). Effects of pH adjustment by parawood ash and effluent recycle ratio on the performance of anaerobic baffled reactors treating high sulfate wastewater. *Bioresource Technology*, *99*, 8987-8994.

Shi, Y., Zhao, X. T., Zhang, Y. M. & Ren, NQ. (2009). Back propagation neural network (BPNN) prediction model and control strategies on methanogen phase reactor treating traditional Chinese medicine wastewater (TCMW). *Journal of Biotechnology*, *144*, 70-74.

Singh, S., Haberl, R., Moog, O., Raj, Shrestha, R., Shrestha, P. & Shrestha, R. (2009). Performance of anaerobic baffled reactor and hybrid constructed wetland treating high-strength wastewater in Nepal-A model for DEWATS. *Journal of Ecological Engineering*, *35*, 654-660.

Skiadas, I. V., Gavala, H. N. & Lyberatos, G. (2000). Modelling of the periodic anaerobic baffled reactor (PABR) based on the retaining factor concept. *Water Research*, *34*(15), 3725-3736.

Song, L. J., Zhu, N. W., Yuan, H. P., Hong, Y. & Ding, J. (2010). Enhancement of waste activated sludge aerobic digestion by electrochemical pre-treatment. *Water Research*, 44 (15), 4371-4378.

Strik, D. P. B. T. B., Domnanovich, A. M., Zani, L., Braun, R. & Holubar, P. (2005). Prediction of trace compounds in biogas from anaerobic digestion using the MATLAB Neural Network Toolbox. *Environmental modeling & Software*, *20*, 803-810.

Tartakovski, B., Mehta, P., Bourque, J. S. & Guiot., S. R. (2011). Electrolysis-enhanced anaerobic digestion of wastewater. *Bioresource Technology*, *102*(10), 5685-5691.

Tawfik, A., Sobhey, M. & Badawy, M. (2008). Treatment of a combined dairy and domestic wastewater in an up-flow anaerobic sludge blanket (UASB) reactor followed by activated sludge (AS system). *Desalination*, *227*, 167-177.

Tchobanoglous, G., Burton, F. L. & Stensel, H. D. (2003). *Wastewater Engineering, Treatment and Reuse*, 4th, edition, Metcalf, and, Eddy, Inc., McGraw-Hill, Inc.

Turkdogan-Aydinol, F. I. & Yetilmezsoy, K. (2010). A fuzzy-logic-based model to predict biogas and methane production rates in a pilot-scale mesophilic UASB reactor treating molasses wastewater. *Journal of Hazardous Materials*, *182*, 460-471.

Uyanik, S., Sallis, P. J. & Anderson, G. K. (2002). The effect of polymer addition on granulation in an anaerobic baffled reactor (ABR). Part I: process performance. *Water Research*, *36*, 933-943.

Voolapalli, R. K. & Stuckey, D. C. (2001). Hydrogen production in anaerobic reactors during shock loads-influence of formate production and H_2 kinetics. *Water Research*, *35*(7), 1831-1841.

Wang J., (2006). *Analytical, Electrochemsitry*, 3rd, edition, Wiley, &, Sons, Inc., New, Jersey, USA.

Wang, J., Huang, Y. & Zhao, X. (2004). Performance and characteristics of an anaerobic baffled reactor. *Bioresource Technology*, *93*(2), 205-208.

Yang, M. & Han-Qing, Y. (2007). Simulation of biological hydrogen production in a UASB reactor using neural network and genetic algorithm. *International journal of Hydrogen Energy*, *32*, 3308-3314.

Yu, H. Q. & Fang, H. H. P. (2003). Acidogenesis of gelatine-rich wastewater in an upflow anaerobic reactor: influene of pH and temperature. *Water Research*, *37*, 55-66.

Yuan, H. P., Yan, X. F., Yang, C. F. & Zhu, N. W. (2011). Enhancement of waste activated sludge dewaterability by electro-chemical pretreatment. *Journal of hazardous materials*, *87* (1-3), 82-88.

Zhu, G. F., Li, J. Z., Wu, P., Jin, H. Z. & Wang, Z. (2008). The performance and phase separated characteristics of an anaerobic baffled reactor treating soybean protein processing wastewater. *Bioresource Technology*, *99*, 8027-8033.

In: Methanogenesis
Editor: Gagik Badalians Gholikandi
ISBN: 978-1-63321-567-2

Chapter 4

MATHEMATICAL MODELING OF ACETOCLASTIC METHANOGENESIS IN TWO-PHASE BIOGAS-PLANTS

Ivo Muha*[*]*, Johannes Schneider and Gabriel Wittum
Goethe-Center for Scientific Computing,
Goethe-University Frankfurt am Main, Germany

Abstract

In this chapter the mathematical modeling of acetoclastic methanogenesis in the second stage, i.e., the anaerobic filter, of a two-phase biogas reactor is investigated. In 2002 a very general model for anaerobic digestion (the ADM1) was published by Batstone et al. (2002). The ADM1 covers all four phases of the anaerobic digestion process (hydrolysis, acidogenesis, acetogenesis and methanogenesis). To get a further understanding of biogas-plants, the process liquid flow within the different reactor parts needs to be considered as well. The main focus of this work is to couple hydraulic patterns of process liquid flow with microbiological processes within one mathematical model. For modeling process liquid flow, the second stage of the biogas reactor is considered to be a porous medium, of which the matrix consists of plastic tower packing and the immobilized microorganisms growing on them. In order to

[*]E-mail address: ivo.muha@gcsc.uni-frankfurt.de

couple flow and microbiological processes the biogas reactor needs to be spatially discretized, meaning that the biogas reactor is approximated by a computational grid consisting of triangles or quadrangles in 2D or prisms in 3D. Typically, in each grid point all variables (concentrations, pressure, etc.) are computed. Furthermore, a new inhibition factor is introduced in order to account for the limited number of plastic tower packing within the reactor. This factor decreases the uptake and growth of microorganisms if the plastic tower packing are being overpopulated. In this chapter the mathematical model is validated by means of experimental data. Furthermore, the following computational results from the derived model are investigated: effectiveness of the anaerobic filter and distribution of anaerobic digestion phases to reactor parts. Finally, it is shown that including process liquid flow within a mathematical model is necessary in order to obtain a reasonable model.

Keywords: Anaerobic digestion, ADM1, Anaerobic filter, Methanogenesis

4.1. Introduction

The capability of obtaining combustible gas with very low impact on the environment plays a key role in longterm sustainable energy supply. In this context, biogas-plants are very important [Chynoweth et al., 2001]. In biogas-plants, biomass is degraded by anaerobic digestion through a cascade of biochemical reactions and is finally converted into biogas, which mainly consists of methane and carbon dioxide. After a clarification step biogas can be inducted into the natural gas net. Another possible application is to fuel a combined heat and power plant for power and heat generation.

In order to develop efficient and sustainable biogas technology, the whole process from biomass production to the construction of biogas-plants needs to be investigated. For example, [Amon et al., 2006a] explored sustainable crop rotations in the production of renewable resources and [Amon et al., 2006b] investigated the influence of biomass composition on methane yield. Optimization of the design of biogas-plants was studied by [Wett et al., 2007]. So called two-phase biogas plants have been studied in several publications because of their higher methane concentration, e.g., [Zielonka et al., 2010, Schönberg and Linke, 2009, Schönberg and Linke, 2012].

Another approach to improve biogas technology is to optimize biogas production by getting a further understanding of the anaerobic digestion process within the biogas-plant. This can be achieved by mathematical modeling.

Within the biogas research community the ADM1 is usually used for modeling the anaerobic digestion [Batstone et al., 2002, Batstone, 2006]. For example, [Boubaker and Ridha, 2008] investigated the anaerobic co-digestion of olive mill wastewater with olive mill solid waste, [Wichern et al., 2009] examined the monofermentation of grass silage under mesophilic conditions, and [Thamsiriroj and Murphy, 2011] addressed the mono-digestion of grass silage in a two-stage CSTR anaerobic digester.

However, one of the major drawbacks of the ADM1 is the lack of consideration of spatial dependencies for the considered substances. In other scientific communities mathematical models considering spatial dependencies, e.g., by modeling a multiphase flow, are widely used. In [Nägel et al., 2008] a diffusion model is used in order to describe the drug transport through human stratum corneum. Another area is hydrogeology where [Grillo et al., 2010] used a model for seepage flow through a porous medium for the simulation of density driven flow in fractured porous media.

The main focus of this work is to show how a model for multi-phase flow can be coupled with the ADM1 in order to include spatial dependencies for the considered substances into the model. One important part of this procedure is the discretization of the computational domain, i.e., the reactor. To this end, the computational domain needs to be covered with geometrical elements such as tetrahedra or prisms (in 3D). Then, the values for all considered substances are computed in each node of the computational grid. This provides additional information which can be used to investigate the distribution of a substance's concentration within the reactor, or to identify ineffective areas of the reactor. As a consequence the whole system can be optimized.

The structure of this chapter is as follows: In Section 4.2. the procedure of coupling a mathematical model for multiphase flow with the ADM1 is presented. Furthermore, the operation of the lab-scale biogas-plant is described. Section 4.3. deals with computational methods and gives detailed information on numerical methods, software tools and the computational grid. In Section 4.4. results are presented. Especially, Section 4.4.4. investigates whether spatial dependencies need to be included into the mathematical model for the considered experimental setup. Finally, Section 4.5. gives some conclusions and an outlook on prospective research subjects.

4.2. Mathematical Model for Process Liquid Flow and Acetoclastic Methanogenesis

In the first part of this section a general mathematical model describing a multi-phase seepage flow through a porous medium is presented. The operation of the considered labscale two-phase biogas-plant is described in the second part. Finally, in the third part of this section the procedure of coupling a multi-phase flow model with a model for acetoclastic methanogensis (based on the ADM1) for the considered biogas-plant is presented.

More general information about coupling a multi-phase flow with the ADM1 can be found in [Muha, 2012].

4.2.1. Modeling of Multi-Phase Flow

The structure of this section is as follows: In the first part the most important notation is introduced and in the second part the conservation equations are presented. Hereby, the multi-phase flow is modeled as a seepage flow through a porous medium, where the solid phase consists of microorganisms and filling material. More information on modeling a multi-phase flow can be found in [Bear, 1972, Bear and Bachmat, 1990, Hassanizadeh and Gray, 1979a, Hassanizadeh and Gray, 1979b] and [Hassanizadeh and Gray, 1980].

4.2.1.1. Notation

This section introduces the notation within the model for multi-phase flow. By considering a general Representative Volume Element (RVE) within the domain, i.e., the biogas reactor, three different states of aggregation can be generally observed: solid, liquid and gaseous (cf. Figure 1).

Let ϕ be the volume fraction of the pore space, S^l the volume fraction of the liquid space with respect to the pore space and S^g the volume fraction of the gaseous phase with respect to the pore space. Then the following equation

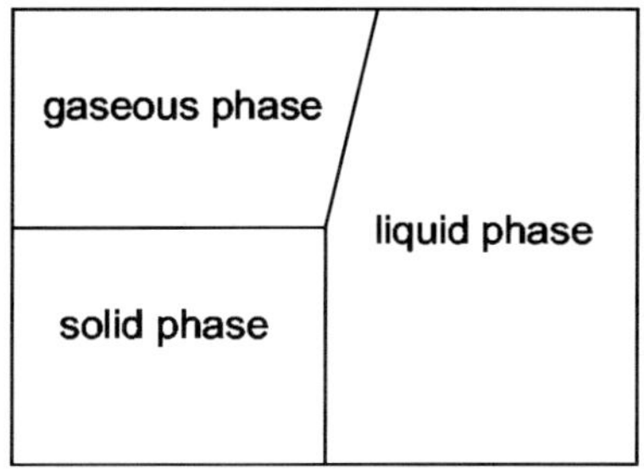

Figure 1. In general, a Representative Volume Element has three different states of aggregation: solid, liquid and gas (redrawn from [Muha, 2012]).

holds

$$S^l + S^g = \phi. \tag{1}$$

In general each phase has a different number of components. Let n_s be the number of components in the solid phase, n_l the number of components in the liquid phase and n_g the number of components in the gaseous phase. Then the mass fraction and the true density of each component with respect to its phase are denoted as

$$\omega_i^j \qquad \text{for } j \in \{s,l,g\} \text{ and } i \in \{1\ldots n_j\} \tag{2}$$

$$\rho_i^j \qquad \text{for } j \in \{s,l,g\} \text{ and } i \in \{1\ldots n_j\}. \tag{3}$$

Note that, definition (2) implies

$$\sum_{i=1}^{n_j} \omega_i^j = 1 \qquad \text{for } j \in \{s,l,g\}. \tag{4}$$

By means of the introduced mass fraction, the phase density ρ^j is computed for each phase as follows:

$$\rho^j = \sum_{i=1}^{n_j} \omega_i^j \rho_i^j \qquad \text{for } j \in \{s,l,g\}. \tag{5}$$

The introduced quantities are sufficient in order to introduce all necessary equations for a multi-phase seepage flow through a porous medium. However, in order to model biochemical reactions additional quantities need to be defined.

Let C_i^j be the concentration of the i-th component within the j-th phase. Then, the concentration is computed as follows:

$$C_i^s = (1-\phi)\rho^s\omega_i^s \tag{6}$$
$$C_i^l = S^l\rho^l\omega_i^l \tag{7}$$
$$C_i^g = S^g\rho^g\omega_i^g. \tag{8}$$

At last, reaction terms are introduced. Let $R_{i_1\to i_2}^{j_1\to j_2}$ be the mass exchange rate of the i_1-th component in the j_1-th phase to the i_2-th component in the j_2-th phase and R_i^j be the total mass exchange rate of the i-th component in the j-th phase. Then, R_i^j is evaluated as follows

$$R_i^j = \sum_{k_1\in\{s,l,g\}} \sum_{k_2=1}^{n_{k_1}} R_{i\to k_2}^{j\to k_1} + R_{k_2\to i}^{k_1\to j}. \tag{9}$$

The mass exchange rate of the j_1-th phase to the j_2-th phase $R^{j_1\to j_2}$ is then given by

$$R^{j_1\to j_2} = \sum_{i_2=1}^{n_{j_2}} \sum_{i_1=1}^{n_{j_1}} R_{i_1\to i_2}^{j_1\to j_2}, \tag{10}$$

and the total mass exchange rate of the j-th phase R^j is derived as

$$R^j = \sum_{j_1\in\{s,l,g\}} R^{j_1\to j} + R^{j\to j_1}. \tag{11}$$

Remark 1. *Total mass of the system is locally conserved, if and only if the following holds*

$$\sum_{j_1\in\{s,l,g\}} \sum_{j_2\in\{s,l,g\}} \sum_{i_1=1}^{n_{j_1}} \sum_{i_2=1}^{n_{j_2}} R_{i_1\to i_2}^{j_1\to j_2} = 0. \tag{12}$$

Table 1. Nomenclature for the most important quantities within the mathematical model with $i_s \in [1,\ldots,n_s]$, $i_l \in [1,\ldots,n_l]$, $i_g \in [1,\ldots,n_g]$, $j_1, j_2, j \in \{s,l,g\}$ und $i_1 \in [1,\ldots,n_{j_1}], i_2 \in [1,\ldots,n_{j_2}], i \in [1,\ldots,n_j]$

Quantity	Describtion
n_s	Number of components within the solid phase
n_l	Number of components within the liquid phase
n_g	Number of components within the gaseous phase
ϕ	Volume fraction of the pore space
S^l	Volume fraction of the liquid space with respect to the pore space
S^g	Volume fraction of the gaseous space with respect to the pore space
$\omega^s_{i_s}$	Mass fraction of the i_s-th component within the solid phase
$\omega^l_{i_l}$	Mass fraction of the i_l-th component within the liquid phase
$\omega^g_{i_g}$	Mass fraction of the i_g-th component within the gaseous phase
$\rho^s_{i_s}$	True density of the i_s-th component within the solid phase
$\rho^l_{i_l}$	True density of the i_l-th component within the liquid phase
$\rho^g_{i_g}$	True density of the i_g-th component within the gaseous phase
ρ^s	Density of the solid phase
ρ^l	Density of the liquid phase
ρ^g	Density of the gaseous phase
$C^s_{i_s}$	Concentration of the i_s-th component within the solid phase
$C^l_{i_l}$	Concentration of the i_l-th component within the liquid phase
$C^g_{i_g}$	Concentration of the i_g-th component within the gaseous phase
$R^{j_1\to j_2}_{i_1\to i_2}$	Exchange of mass from i_1-th component of the j_1-th phase to the i_2-th component of the j_2-th phase
$R^{j_1\to j_2}$	Exchange of mass from the j_1-th phase to the j_2-th phase
R^j_i	Sum of all reaction terms of the i-th component within the j-th phase
R^j	Sum of all reaction terms of the j-th phase

4.2.1.2. Mass Conservation Equations

After the notation was introduced in the previous section, the PDEs describing mass-conservation equations for each component and each phase will be stated in this section.

The mass conservation equation for the solid phase reads

$$\frac{\partial((1-\phi)\rho^s)}{\partial t} = R^s, \tag{13}$$

and the mass conservation equation for each component within the solid phase reads

$$\frac{\partial((1-\phi)\rho^s\omega_i^s)}{\partial t} = R_i^s \qquad \text{for } i \in [1\ldots n_s]. \tag{14}$$

Remark 2. *Note that, the volume fraction of the solid phase ϕ is not constant with respect to space and time. This is necessary in order to be able to model the attachment and detachment of microorganisms to the filling material.*

In order to introduce the mass conservation equation for the liquid phase, its Darcy velocity $\mathbf{q}_l$ needs to be introduced first

$$\mathbf{q}_l = -\frac{\mathbb{K}_l}{\mu_l}(\nabla p_l - \rho^l \mathbf{g}). \tag{15}$$

Hereby, $\mathbb{K}_l$ is the permeability tensor and μ_l is viscosity. The Darcy velocity $\mathbf{q}_l$ describes the velocity of the liquid phase, which is driven by pressure and gravity.

Finally, the mass conservation equation for the liquid phase reads

$$\frac{\partial(\rho^l S^l)}{\partial t} = -\nabla(\rho^l \mathbf{q}_l) + R^l. \tag{16}$$

Within the liquid phase each component is additionally flowing by means of diffusion, which is described by the diffusion tensor $\mathbb{D}_i^l$. Then, the mass conservation equation is stated as

$$\frac{\partial(\rho^l S^l \omega_i^l)}{\partial t} = -\nabla(\rho^l \mathbf{q}_l \omega_i^l - \rho^l \mathbb{D}_i^l \nabla \omega_i^l) + R_i^l \qquad \text{for } i \in [1\ldots n_l]. \tag{17}$$

The equations for the gas phase are derived analogously as:

$$\frac{\partial(\rho^g S^g)}{\partial t} = -\nabla(\rho^g \mathbf{q}_g) + R^g \tag{18}$$

$$\mathbf{q}_g = -\frac{\mathbb{K}_g}{\mu_g}(\nabla p^g - \rho^g \mathbf{g}) \tag{19}$$

$$\frac{\partial(\rho^g S^g \omega_i^g)}{\partial t} = -\nabla(\rho^g \mathbf{q}_g \omega_i^g - \rho^g \mathbb{D}_i^g \nabla \omega_i^g) + R_i^g \qquad \text{for } i \in [1 \ldots n_g]. \tag{20}$$

However, in contrary to p^l, which is considered to be an unknown, p^g is usually given as a function of p_l and S_l.

$$p^g = p^g(p^l, S^l). \tag{21}$$

4.2.2. Operation of the Labscale Two-phase Biogas-Plant

In order to decide which part of the ADM1 needs to be included within the final mathematical model, the operation of the considered labscale biogas-plant is investigated first. The structure of this section is as follows. Section 4.2.2.1. briefly describes the experimental setup of the lab-scale biogas-plant and Section 4.2.2.2. gives details about the operation of the biogas-plant.

4.2.2.1. Experimental Setup

The experimental setup is shown in Figure 2. The biogas reactor consists of three major parts. The first part is the hydrolysis reactor Figure 2 [1]. It has an effective volume of $100l$ and its operating temperature is $55°C$. The second part is a process liquid reservoir Figure 2 [2] with an effective volume of approximately $55l$. The third part is the anaerobic filter, which is a downstream fixed film cylindric reactor operated at $38°C$. Its height is $0.71m$ and its diameter is $0.24m$. It has a total usable volume of approximately $32l$. The main focus of this work lies on this part of the biogas-plant. Furthermore, the reactor is equipped with an automated gas analysis system Figure 2 [7].

The process liquid circulates by two independent circuits. In the primary circuit Figure 2 [4], the process liquid circulates from the reservoir to the hydrolysis reactor with a total flow of $60lh^{-1}$. In the secondary circulation system

Figure 2 [5], the liquid is transferred from the reservoir to the anaerobic filter and then to the hydrolysis reactor with a total flow of $1lh^{-1}$. Further information on this reactor can be found in [Schönberg and Linke, 2009] and [Schönberg and Linke, 2012].

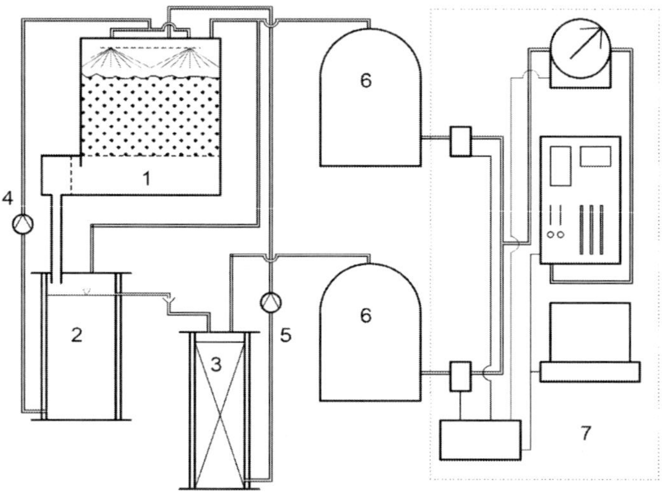

Figure 2. Scheme of a two-phase laboratory scale anaerobic digestion system consisting of: [1] a hydrolysis reactor ($100l$), [2] a reservoir for leachate from the hydrolysis reactor ($60l$), [3] a fixed film anaerobic filter ($32.12l$), [4] a small leachate circulation system ($60lh^{-1}$), [5] a complete leachate circulation system ($1lh^{-1}$) [6] two biogas bags and [7] an automated gas analysis system (redrawn from [Muha et al., 2012]).

4.2.2.2. Experimental Procedure

Each experiment had a total length of 21 days. In the beginning of each experiment, the hydrolysis reactor was fed with a total of 11kg biomass (10kg rye silage, 1kg rye straw). Due to the primary circulation system, process liquid was led through the hydrolysis reactor and therefore, organic compounds accumulated within the process liquid. The liquid was then transferred to the reservoir and by means of the secondary circulation system transferred even further to the anaerobic filter. There the organic compounds were degraded and biogas formation occurred. The process liquid was analyzed each day at two different points within the secondary circulation system. The first point was before entering the anaerobic filter and the second point was after leaving the anaerobic filter. The collected quantities, which are relevant for this article, are shown in Tab. 2. Additionally, methane and biogas yield from the anaerobic filter were measured by means of the automated gas analysis system.

Table 2. Some measured quantities within the secondary circulation system

Quantity	Measured before the anaerobic filter	Measured after the anaerobic filter
pH	yes	yes
acetic acid	yes	no
propionic acid	yes	no
butyric acid	yes	no
valeric acid	yes	no
caproic acid	yes	no
acetic acid equivalence	yes	yes

4.2.3. Coupling of a Model for Acetoclastic Methanogenesis with the Model for Multi-Phase Flow

The ADM1 was published in 2002 [Batstone et al., 2002] and considers a total of 17 substances and six different species of microorganisms. In total it consists of 24 ordinary differential equations and some algebraic equations. In this section we will show how these ordinary differential equations can be combined with the partial differential equations of the multi-phase-flow model with respect to the considered two-phase biogas-plant from Section 4.2.2.1..

First, the constituents to be considered need to be specified. Since the mathematical model needs to be validated in the end, this has to be performed considering the measured quantities from the experiment. We choose to consider the following constituents:

1. filling material
2. microorganisms
3. acetic acid equivalence
4. water
5. methane
6. carbon dioxide

The next step is to assign a state of aggregation to each constituent: We choose to consider the following substances within the solid phase $\mathcal{P}_s$:

$$\mathcal{P}_s := \{acd, PTP\}, \tag{22}$$

where acd represents microorganisms and PTP the plastic tower packing, which is the filling material of the anaerobic filter.

According to Section 4.2.1.1. we introduce the following notation: The mass fraction of the microorganisms is denoted by ω^s_{acd} and the mass fraction of the plastic tower packing by ω^s_{PTP} . The apparent density of the solid phase ρ_s is

$$\rho_s = \omega^s_{acd}\rho^s_{acd} + (1-\omega^s_{acd})\rho^s_{PTP}, \tag{23}$$

where ρ^s_{acd} is the density of the microorganisms and ρ^s_{PTP} the density of the plastic tower packing. Note that the sum of all mass fractions within a single phase equals one. Therefore, the equation $(1-\omega^s_{acd}) = \omega^s_{PTP}$ holds.

For the liquid phase $\mathcal{P}_l$ we consider

$$\mathcal{P}_l := \{H_2O, ac\}. \tag{24}$$

Hereby, H_2O represents water and ac represents acetic acid equivalence.

Then, analogously to the notation of the solid phase, $\omega^l_{H_2O}$ denotes the mass fraction of water and ω^l_{ac} the mass fraction of acetic acid equivalence. Again as stated in Section 4.2.1.1., the following equations hold

$$\omega^l_{H_2O} = 1-\omega^l_{ac} \tag{25}$$
$$\rho_l = \omega^l_{ac}\rho^l_{ac} + (1-\omega^l_{ac})\rho^l_{H_2O}, \tag{26}$$

where ρ^l_{ac} is the density of acetic acid equivalence and $\rho^l_{H_2O}$ is the density of water.

At last, the gas phase $\mathcal{P}_g$ is defined as follows:

$$\mathcal{P}_g := \{CH_4, CO_2\}, \tag{27}$$

where CH_4 represents methane and CO_2 represents carbon dioxide. Again, $\omega^g_{CH_4}$ denotes the mass fraction of methane and $\omega^g_{CO_2}$ the mass fraction of carbon dioxide. It holds that:

$$\omega^g_{CO_2} = 1 - \omega^g_{CH_4} \tag{28}$$

$$\rho_g = \omega^g_{CH_4}\rho^g_{CH_4} + (1 - \omega^g_{CH_4})\rho^l_{CO_2}. \tag{29}$$

Remark 3. *Note that, we do not consider the solubility of carbon dioxide. However, if necessary carbonic acid can be added to the liquid phase.*

After introducing all constituents that ought to be considered, the reaction terms $R^{j_1 \to j_2}_{i_1 \to i_2}$ still need to be specified (cf. Section 4.2.1.2.). We take these reaction terms from the corresponding ordinary differential equations of the ADM1. Using this rationale the non-zero reaction terms are listed in Tab. 3 and they are stated as follows:

Table 3. Denotation of reaction terms included into the mathematical model

Denotation	Process
$R^{l \to s}_{ac \to acd}$	consumption of acetate by microorganisms (proliferation)
$R^{l \to g}_{ac \to CH_4}$	degradation of acetate to methane
$R^{l \to g}_{ac \to CO_2}$	degradation of acetate to carbon dioxide
$R^{s \to l}_{acd \to H_2O}$	transfer of microorganisms to the liquid phase (death)
$R^{s \to l}_{acd \to ac}$	transfer of microorganisms to acetate (death)
$R^{l \to s}$	phase transition (from liquid to solid phase)
$R^{s \to l}$	phase transition (from solid to liquid phase)
$R^{l \to g}$	phase transition (from liquid to gas phase)

Consumption of acetate by microorganisms

$$R^{l\to s}_{ac\to acd} = k_{m,acd} Y_{acd} \frac{C^l_{ac}}{K_{s,ac} + C^l_{ac}} C^s_{acd} I_{acd} \tag{30}$$

Degradation of acetate to methane

$$R^{l\to g}_{ac\to CH_4} = \frac{4}{15} k_{m,acd} (1 - Y_{acd}) \frac{C^l_{ac}}{K_{s,ac} + C^l_{ac}} C^s_{acd} I_{acd} \tag{31}$$

Degradation of acetate to carbon dioxide

$$R^{l\to g}_{ac\to CO_2} = \frac{11}{15} k_{m,acd} (1 - Y_{acd}) \frac{C^l_{ac}}{K_{s,ac} + C^l_{ac}} C^s_{acd} I_{acd} \tag{32}$$

The fractions $4/15$ and $11/15$ need to be introduced because of the conversion of units (cf. Section 4.2.3.1.).

Death of microorganisms

$$R^{s\to l}_{acd\to ac} = k_{dec,acd} f_{ac,acd} C^s_{acd} \tag{33}$$

$$R^{s\to l}_{acd\to H_2O} = k_{dec,acd} (1 - f_{ac,acd}) C^s_{acd} \tag{34}$$

In contrary to the ADM1 model the multi-phase model includes spatial dependencies. Therefore, we use a different inhibition term:

$$I_{acd} = I^{MAX}_{acd} = 1 - \frac{C^s_{acd}}{acd_{MAX,acd}} \tag{35}$$

This inhibition term reduces uptake of acetate if the local concentration of microorganisms is to high ($C^s_{acd} \approx acd_{MAX,acd}$), which is the case when the filling material is being overpopulated. On the other hand, if local concentration of microorganisms is low ($C^s_{acd} \approx 0$) then this inhibition term has no impact on reaction kinetics, meaning that, uptake is linear with respect to C^s_{acd}.

4.2.3.1. Conversion of units

In the ADM1, COD concentration is used as default unit. On the other hand the default unit in multi-phase flow models is density. Therefore, all parameters need to be converted to the new default units. This can be performed by means of the following equation:

$$C^{kg/l} = C^{kgCOD/m^3} \cdot 1/\mathrm{Co}^{COD} \cdot M \cdot \frac{m^3}{1000l}, \tag{36}$$

where Co^{COD} describes the COD content ($[\mathrm{Co}^{COD}] = \mathrm{kgCOD/mol}$) and M the molar mass ($[M] = \mathrm{kg/mol}$) of the considered substance.

4.2.3.2. Simplifications of the Model

The model as it is stated in the previous section also includes the flow of the gaseous phase within the anaerobic filter. In order to keep computational cost at an acceptable level, the model is simplified by assuming that:

- The produced gas is removed immediately from the reactor.
- The volume of the gas phase is zero.
- The gas has no influence on kinetics.

With these assumptions at hand we can set $S^g = 0$ and, finally, derive the following set of equations:

$$\frac{\partial((1-\phi)\rho^s)}{\partial t} = R^s \tag{37}$$

$$\frac{\partial((1-\phi)\rho^s\omega^s_{acd})}{\partial t} = R^s_{acd} \tag{38}$$

$$\frac{\partial(\phi\rho^l)}{\partial t} = -\nabla(\rho^l \mathbf{q}_l) + R^l \tag{39}$$

$$\frac{\partial(\phi\rho^l\omega^l_{ac})}{\partial t} = -\nabla(\rho^l \mathbf{q}_l \omega^l_{ac} - \rho^l \mathbb{D}^l_{ac}(\phi)\nabla\omega^l_{ac}) + R^l_{ac} \tag{40}$$

$$P^g_{CH_4}(t) = \int_{T_0}^{t} R^g_{CH_4}(\hat{t})d\hat{t} \tag{41}$$

$$P^g_{CO_2}(t) = \int_{T_0}^{t} R^g_{CO_2}(\hat{t})d\hat{t} \tag{42}$$

with (according to Section 4.2.1.1. und 4.2.1.2.)

$$\mathbf{q}_l = -\frac{\mathbb{K}_l}{\mu_l}(\nabla p_l - \rho^l \mathbf{g}) \tag{43}$$

$$\rho^s = \rho^s_{acd}\omega^s_{acd} + (1-\omega^s_{acd})\rho^s_{PTP} \tag{44}$$

$$\rho^l = \rho^l_{ac}\omega^l_{ac} + (1-\omega^l_{ac})\rho^s_{H_2O} \tag{45}$$

$$R^s_{acd} = R^{l\to s}_{ac\to acd} - R^{s\to l}_{acd\to ac} - R^{s\to l}_{acd\to H_2O} \tag{46}$$

$$R^l_{ac} = R^{s\to l}_{acd\to ac} - R^{l\to s}_{ac\to acd} - R^{l\to g}_{ac\to CH_4} - R^{l\to g}_{ac\to CO_2} \tag{47}$$

$$R^g_{CH_4} = R^{l\to g}_{ac\to CH_4} \tag{48}$$

$$R^g_{CO_2} = R^{l\to g}_{ac\to CO_2} \tag{49}$$

$$R^s = R^s_{acd} \tag{50}$$

$$R^l = -R^s_{acd} - R^g_{CH_4} - R^g_{CO_2}. \tag{51}$$

Remark 4. *Note that, the gas production is now measured by means of a total production $P^g_{CH_4}(t)$ and $P^g_{CO_2}(t)$ instead of using a mass fraction.*

The parameter used in the model are explained in Section 4.2.3.3.

4.2.3.3. Model parameters

In this section all model parameters are defined. The parameters can be split into two groups. The first group arises from the use of a multi-phase flow model and is shown in Tab. 4. The second group of parameters is introduced by means of the ADM1 and is shown in Tab. 5. The parameter values used within simulations can be found in Tab. 6. Hereby, the second group of parameters was estimated by comparing simulation results with experiments. As parameter estimation techniques are not the main focus of this work no details are provided. However, they can be found in [Muha, 2012].

4.3. Computational Methods

This section describes the computational methods. In Section 4.3.1. the applied numerical methods are stated. In Section 4.3.2. the utilized software tools are listed and Section 4.3.3. gives details on the numerical setup, i.e., the computational grid, boundary and initial conditions.

Table 4. Multi-phase flow parameters

Parameter	Description
ρ^s_{PTP}	true density of filling material
ρ^s_{acd}	true density of microorganisms
$\rho^l_{H_2O}$	true density of water
ρ^l_{ac}	true density of acetic acid
$\mathbb{K}_l$	permeability
μ_l	viscosity of the liquid phase
$\mathbb{D}^l_{ac}$	diffusion tensor of acetic acid in water
$\mathbf{g}$	gravity acceleration vector

Table 5. Kinetic parameters from the ADM1 model.

Parameter	Description
$k_{m,ac}$	uptake rate of microorganisms
Y_{ac}	fraction of uptaken acetate used for growth
$K_{s,ac}$	Michaelis-Menten constant for acetate
$f_{ac,acd}$	fraction of decayed microorganisms converted to acetate
$k_{dec,acd}$	decay rate of microorganisms
$acd_{MAX,acd}$	plastic tower packing dependent parameter

4.3.1. Numerical Methods

In order to solve the derived partial differential equations (37)-(51) several numerical methods need to be applied.

The discretization in time is performed by means of a semi-implicit Euler scheme, the discretization in space is performed by a Finite Volume Method (with upwind). As the equations are non-linear Newton's method is used in each time-step. Due to the discretization in space, the linear system has a very large number of degrees of freedom (up to $1 \cdot 10^5$). Therefore, a geometric multigrid solver is applied to solve the system of linear equations.

More information about these methods can be found in [Ortega and

Table 6. Fixed model parameters and reaction parameters estimated according to [Muha et al., 2012]

Fixed parameter	Value	Unit	Est. parameter	Value	Unit
ρ^s_{acd}	1090	$kg \cdot m^{-3}$	$k_{\mathrm{m,ac}}$	$1.54 \cdot 10^{-2}$	h^{-1}
ρ^s_{PTP}	950	$kg \cdot m^{-3}$	Y_{ac}	$1.88 \cdot 10^{-1}$	1
ρ^l_{ac}	1050	$kg \cdot m^{-3}$	$K_{S_{ac}}$	$2.00 \cdot 10^{0}$	$kg \cdot m^{-3}$
$\rho^l_{H_2O}$	1000	$kg \cdot m^{-3}$	k_{dec}	$4.16 \cdot 10^{-4}$	h^{-1}
$\mathbb{D}^l_{ac}$	$1.210 \cdot 10^{-9} \cdot \mathbb{1}$	$m^2 \cdot s^{-1}$	$f_{ac,acd}$	$1.00 \cdot 10^{0}$	1
$\mathbf{g}$	$(0, -9.806)$	$m \cdot s^{-2}$	acd_{max}	$5.47 \cdot 10^{1}$	$kg \cdot m^{-3}$

Rheinboldt, 1987, Deuflhard and Bornemann, 2002, Deuflhard and Hohmann, 1991, Dziuk, 2010, Gilbarg and Trudinger, 1998, Bey, 1997].

Remark 5. *Within the ADM1, the spatial dependencies are neglected. Therefore, the system of linear equations, which need to be solved in every time step, is small and can be solved by any ordinary solver for linear systems.*

4.3.2. Software Tools

In order to perform the computations two software tools were used: ProMesh and UG4. Details are provided below.

4.3.2.1. ProMesh

ProMesh was developed at the Goethe-Center for Scientific Computing by Sebastian Reiter. It is capable of creating and optimizing computational grids in 1D, 2D and 3D. More information can be found in [Reiter and Wittum, 2014].

4.3.2.2. UG4

UG4 is a software tool for robust and efficient solving of partial differential equations. It has lots of different discretization methods and numerical solvers built-in. Furthermore, it is heavily parallelized and can thus be used in parallel on high performance computers. More information on the development of UG and the most recent version, UG4, can be found in [Bastian and Wittum, 1994, Lemke et al., 2012, Vogel et al., 2012].

4.3.3. Numerical Setup

This section covers all details about the computational grid as well as boundary and initial conditions.

4.3.3.1. Computational Grid

In total three different computational grids are used. The first computational grid is 1D and consists of a line along the z-axis. As the anaerobic filter is of cylindric shape this computational grid can only be used if the problem is axial-symmetric and homogenous along the r-axis.
The second computational grid is 2D with coordinates (r,z). Therefore, it can be used for axial-symmetric problems.
The last computational grid is 3D with coordinates (x,y,z). This grid can be used for any kind of problem. The 2D and 3D computational grids are depicted in Fig. 3. Note that, the grid used within the simulations has a higher level of refinement (more gridpoints).

4.3.3.2. Boundary conditions and initial conditions

Linear (w.r.t the z-axis) initial conditions are used for the following unknowns: ω^s_{acd}, ϕ and ω^l_{ac}. The functions are chosen in a way to reflect the initial state of the anaerobic filter as good as possible. The liquid pressure p_l is chosen to be a linear function considering hydrostatic initial conditions.

As the solid phase cannot be transferred out or in the reactor, boundary conditions are not used. Because of the assumptions in Section 4.2.3.2. this also holds for the gas phase. For the liquid-phase and ω^l_{ac} we use the following boundary conditions:

- Top plate: Boundary conditions according to the experiments. They can be incorporated by means of

$$\mathbf{q}_l \cdot \mathbf{n} = \frac{11}{1\text{h}(0.12\,\text{m})^2\pi} \tag{52}$$

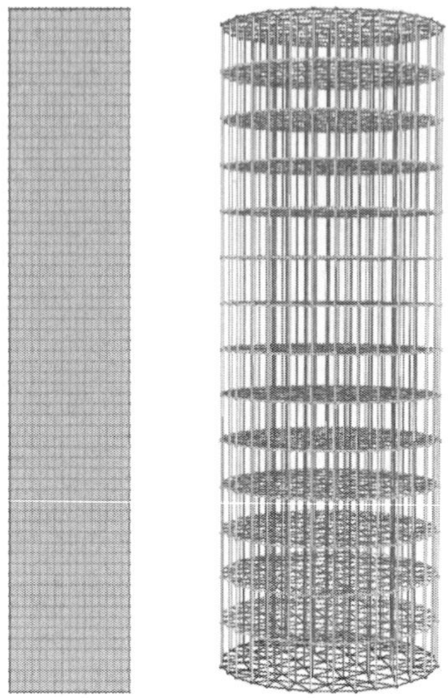

Figure 3. Computational grid: 2D and 3D.

- Cylinder jacket: Zero-flux (impervious wall) boundary conditions are used in order to model the impermeability of the cylinder jacket.
- Bottom plate: Dirichlet zero boundary conditions are used in order to measure the concentration of acetic acid equivalence in the outflowing process liquid.

4.4. Results

In this section results are presented. In Section 4.4.1. it is shown that the mathematical model is capable of simulating acetoclastic methanogenesis in two phase biogas-plants. Then, Section 4.4.2. deals with the efficiency of a biogas-plant and in Section 4.4.3. the distribution of the biological phases to reactor parts is investigated. Some of these results have already been published in [Muha et al., 2012] and [Muha et al., 2013]. Finally, Section 4.4.4. shows that taking spatial dependencies, e.g., by using PDEs, into account is necessary in order to obtain a convincing mathematical model.

4.4.1. Validation of the Mathematical Model

The mathematical model was validated by comparing simulational data with experimental data. In total five experiments (cf. Section 4.2.2.) were investigated. The measured concentration of acetic acid equivalence in the outflowing process liquid was compared to the corresponding values obtained by simulations. The computed acetic acid equivalence concentration is similar to experimental data (cf. Figure 5) by using the same set of parameters. More details can be found in [Muha et al., 2012].

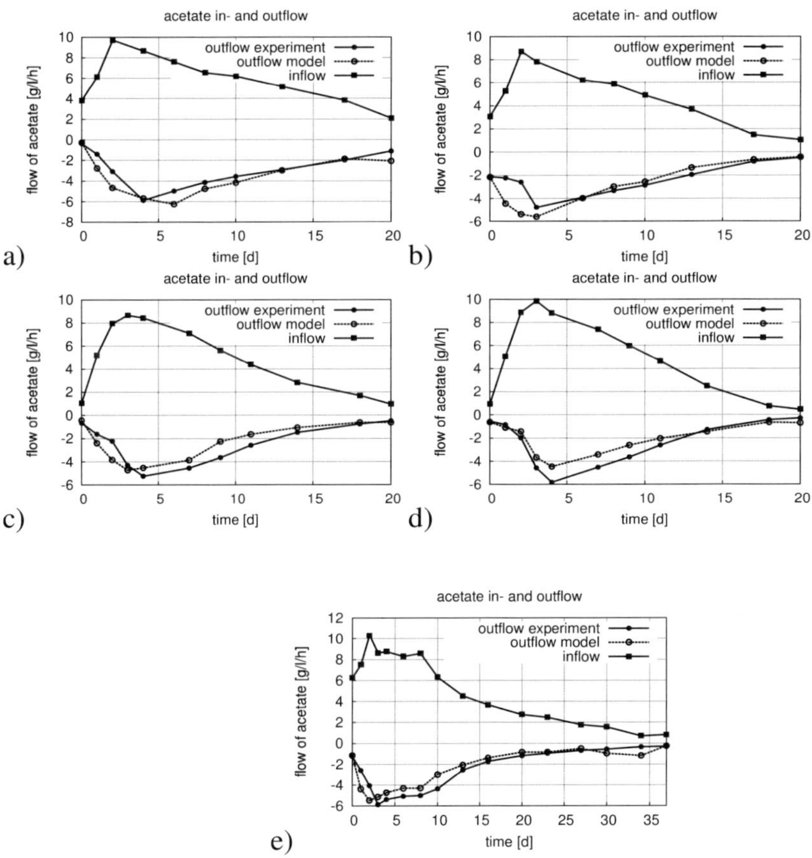

Figure 4. Comparison of model outflow of acetate with experimental data for the five considered model cases (redrawn from [Muha et al., 2012]).

4.4.2. Effectivity of the Anaerobic Filter

In order to investigate the effectivity of the anaerobic filter, the maximal local methane production is defined:

$$P_{\max}(t) := \max_{\mathbf{x}} P_{CH_4}(t,\mathbf{x}). \tag{53}$$

By means of $P_{\max}(t)$ the effectiveness of the anaerobic filter is then given by

$$\mathrm{Eff}(t) := \frac{1}{V_\mathrm{P}} \int_{V_\mathrm{P}} \frac{P_{CH_4}(t,\mathbf{x})}{P_{\max}(t)} dx, \tag{54}$$

with V_P being the porespace.

For an anaerobic filter, which operates at a high effectivity $(\mathrm{Eff}(t) \approx 1)$, a similar amount of methane/biogas is produced everywhere within the reactor. On the other hand, a low effectivity $(\mathrm{Eff}(t) \ll 1)$ means that there is no methane/biogas formation in some parts of the reactor.

In Figure 5 the fraction $P_{CH_4}(t,\mathbf{x})/P_{\max}(t)$ is plotted for different points in time within one experiment. It can be observed that the reactor is not used in an effective way in the experiment. Especially in the beginning and in the end, only the top part of the reactor is used for methane/biogas formation. This is due to the fact that in the beginning and in the end of each experiment, the concentration of acetic acid equivalence within the inflowing process liquid is not high enough and is thus consumed by microorganisms in the top part of the anaerobic filter. Only in the middle of each experiment ($t \approx 12d$) enough acetic acid equivalence is provided to feed the reactor in an effective way. More details can be found in [Muha et al., 2012] and [Muha, 2012].

4.4.3. Distribution of Anaerobic Digestion Phases to Reactor Parts

Anaerobic digestion consists of four consecutive phases:

- hydrolysis
- acidogenesis
- acetogenesis
- methanogensis

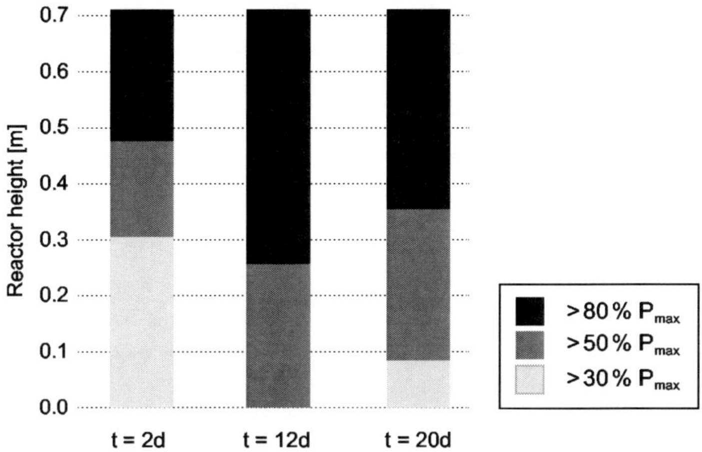

Figure 5. Efficiency of the anaerobic filter (redrawn from [Muha et al., 2012]).

The main idea of two phase biogas-plants is to separate these anaerobic digestion phases into different reactor parts. For example, by restricting the methanogenesis phase to the anaerobic filter methane formation would primarily occur in the anaerobic filter, while on the other hand carbon dioxide formation would also occur in the hydrolysis stage. This would lead to higher methane concentrations in the gas of the anaerobic filter. In [Muha et al., 2013] the mathematical model from Section 4.2. was used and extended in order to compute the distribution of biological phases to reactor parts. Some results are shown in Tab. 7. It can be observed that about 74% of the methanogenesis occurs in the anaerobic filter. Therefore, it seems possible to restrict a large fraction of methanogenesis to a single reactor part. However, restricting the acidogenesis or the acetogenesis to a single reactor part does not work with the considered reactor setup.

4.4.4. Numerical Experiments on Spatial Dependencies

In this section we show that considering process liquid flow in mathematical models for biogas-plants is necessary in order to obtain a convincing model. Therefore, numerical experiments with computational grids in 1D, 2D and 3D considering three different test cases were performed. In all test cases, exper-

Table 7. Distribution of anaerobic digestion phases to reactor parts: HYD: hydrolysis, ACI: acidogenesis, ACE: acetogenesis, MET: methanogenesis, H: hydrolysis reactor, F: anaerobic filter. (taken from [Muha et al., 2013])

Biological Phase	Average
S^H_{HYD}	100 %
S^F_{HYD}	0 %
S^H_{ACI}	60 %± 6 %
S^F_{ACI}	40 %± 6 %
S^H_{ACE}	54 %± 9 %
S^F_{ACE}	46 %± 9 %
S^H_{MET}	26 %± 15 %
S^F_{MET}	74 %± 15 %

iment number two (Figure 4 b) was used and the concentration of acetate at $t = 3$d within the anaerobic filter was investigated.

4.4.4.1. Case 1

In the first case, the whole top plate of the reactor is used homogeneously for process liquid inflow. Therefore, this case was investigated in 1D, 2D and 3D. Results are shown in Figure 6. Regardless of the chosen spatial dimension, all computations yield the same results. However, it is clearly observable that there are large differences in the concentration of acetic acid equivalence along the z-axis. Therefore, even in this simple case a mathematical model ignoring spatial dependencies, i.e., using only ODEs, cannot yield a useful result.

Remark 6. *Note that the 3D plot has a slightly different perspective and therefore might suggest that the result in 3D is different. As already stated, this is not the case.*

4.4.4.2. Case 2

In the second case, the inflow of the process liquid is limited to the center of the top plate (cf. Figure 7). As the area of inflow is decreased, the amount of in-

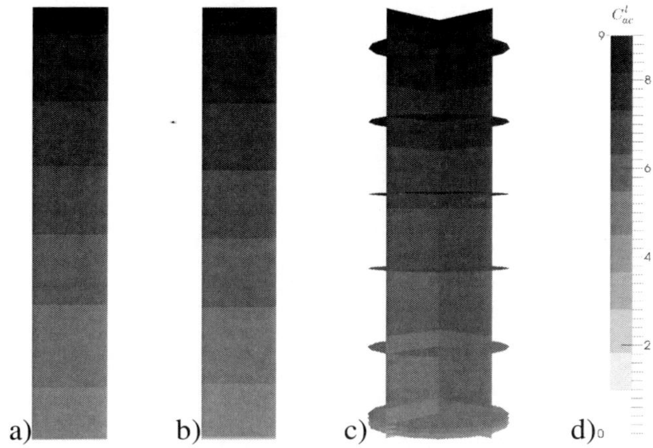

Figure 6. Concentration of acetic acid equivalence in case 1: a) 1D computation (only z-axis), b) 2D computation (cylindric coordinates), c) 3D computation d) legend.

flowing liquid has to be increased in order to conform to the experimental setup ($1l/h$). This numerical setup cannot be incorporated in a 1D computational grid, because the problem is not homogenous along the r-axis. Therefore, this case was investigated in 2D and 3D. Results are shown in Figure 8. Again, as in the first case, the results of the 2D and 3D computation are the same. However, it can be observed that there are differences in the concentration of acetic acid equivalence along the z-axis and r-axis (cylindric coordinates). Therefore, by considering this more realistic case at least a 2D computational grid is necessary in order to determine useful results.

4.4.4.3. Case 3

In the third case the inflow of the process liquid is limited to one part of the top plate on the side (cf. Figure 9). Analogously to case 2, the amount of inflowing liquid had to be increased in order to conform to the experimental setup ($1l/h$). As this numerical setup is not axial-symmetric anymore it cannot be incorporated in a 1D or 2D computational grid. Therefore, it is investigated in 3D only. Results are shown in Figure 10. It can be observed that the concentra-

Figure 7. Case 2: Top plate of the anaerobic filter: grey: used for process liquid inflow, black: no-flow boundary conditions.

tion of acetic acid equivalence varies along the z-axis as well as x- and y- axes. Therefore, in this even more realistic case only 3D computations yield useful results.

4.5. Conclusion

In this chapter, the procedure of coupling a mathematical model for multiphase flow through a porous medium with the ADM1 is presented in detail (Section 4.2.3.). The derived model considers process liquid flow as well as flow of the gaseous phase. However, in order to reduce the computational cost, some reasonable assumptions are made, and the model is reduced (Section 4.2.3.2.) by neglecting the flow of the gaseous phase.

The derived model has already been published in [Muha et al., 2012]. However, the procedure of coupling a mathematical model for multiphase flow and the ADM1 model is new. Section 4.4.1. shows that the presented mathematical model can be validated by means of experimental data and can thus be used for numerical experiments.

Section 4.4.2. and 4.4.3. show some results of the derived model (from [Muha et al., 2012] and [Muha et al., 2013]). These results are very important and are currently being investigated in projects funded by the BMBF.

Finally, Section 4.4.4. investigates the spatial distribution of acetic acid equivalence within the anaerobic filter. It is shown that the assumption of a

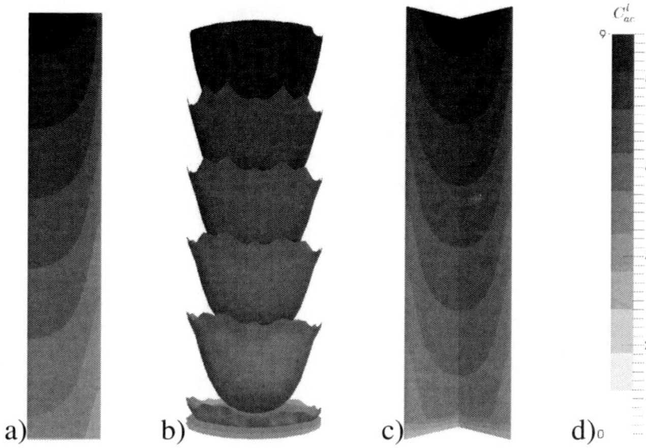

Figure 8. Concentration of acetic acid equivalence in case 2: a) 2D computation (cylindric coordinates), b) 3D computation: iso-surfaces c) 3D computation: slices d) legend.

spatially constant acetic acid equivalence concentration, as it is currently used in ODE based models, is not advisable for the considered reactor setup. Under certain circumstances, i.e., a homogeneous inflow of process liquid in the whole top plate, 1D models can be used. However, this is a strong assumption and should not be used in general.

Therefore, the main result of this work is that, for the considered biogas-plant, at least 2D computations should be used in order to obtain reasonable results.

4.6. Outlook

4.6.1. Increasing the Effectiveness of the Anaerobic Filter

As it is shown in Section 4.4.2. the anaerobic filter is not used very effectively. This is because most of the time the total inflow of acetic acid equivalence is not sufficient. The effectiveness of the anaerobic filter can be increased by one of the following approaches:

- changing reactor geometry, i.e., decreasing reactor height;

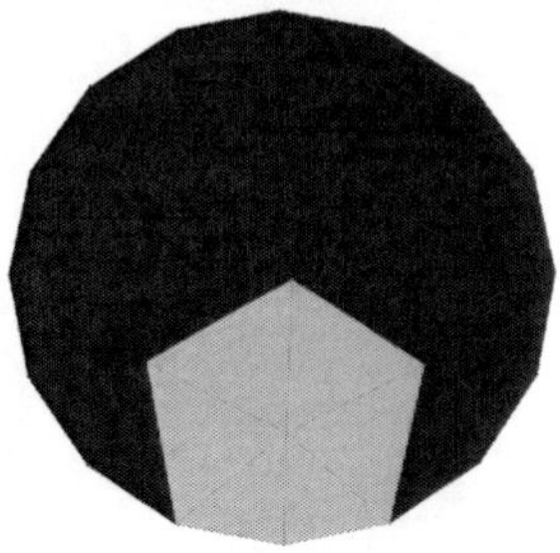

Figure 9. Case 3: Top plate of the anaerobic filter: grey: used for process liquid inflow, black: no-flow boundary conditions.

- adding a process liquid inflow in the middle of the reactor;
- using different filling material in different regions of the anaerobic filter.

The latter point is currently being investigated within the Methanoquant project in cooperation with BTU Cottbus - Senftenberg.

4.6.2. Understanding the Processes within a Biogas-plant

Biogas-plants are often referred to as black boxes. Most of the microorganisms and lots of different biochemical processes are still unknown. In [Muha et al., 2013], a first step in breaking down the black box "biogas-plant" is made by computing the distribution of the anaerobic digestion phases to reactor parts. However, in order to get a deeper understanding of biogas-plants, measuring techniques need to be improved. This is currently being investigated in a cooperation with the BTU Cottbus - Senftenberg and the University of Hohenheim.

4.6.3. Extension of the Mathematical Model

The procedure of coupling a mathematical model is stated in a very general sense. Therefore, this procedure can be used to expand the mathematical model to include other substances and other biological and biochemical phenomena. This is going to be more and more important since measuring techniques improve and more experimental data becomes available.

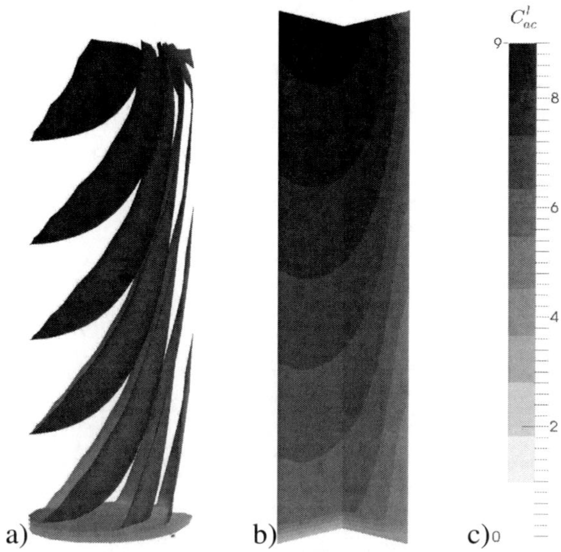

Figure 10. Concentration of acetic acid equivalence in case 3: a) 3D computation: iso-surfaces b) 3D computation: slices c) legend.

Acknowledgments

This research was supported by the German Ministry for Education and Research (BMBF) under the contract number 03SF0423A.

References

[Amon et al., 2006a] Amon, T., Amon, B., Kryvoruchko, V., Machmüller, A., Hopfner-Sixt, K., Bodiroza, V., Hrbek, R., Friedel, J., Pötsch, E., Wagentristl, H., Schreiner, M., and Zollitsch, W. (2006a). Methane production through anaerobic digestion of various energy crops grown in sustainable crop rotations. *Bioresource Technology*, 98:3204–3212.

[Amon et al., 2006b] Amon, T., Amon, B., Kryvoruchko, V., Zollitsch, W., Mayer, K., and Gruber, L. (2006b). Biogas production from maize and dairy cattle manure - Influence of biomass composition on the methane yield. *Agriculture, Ecosystems and Environment*, 118:173–182.

[Bastian and Wittum, 1994] Bastian, P. and Wittum, G. (1994). Adaptive Multigrid Methods: The UG concept. In W. Hackbusch; G. Wittum, editor, *Adaptive Methods - Algorithms, Theory and Applications (Proceedings of the 9th GAMM Seminar Kiel)*, volume 46 of *Notes on Numerical Fluid Mechanics*. Vieweg.

[Batstone, 2006] Batstone, editor (2006). *Anaerobic Digestion Model No 1 - Developments and Applications*. IWA Publishing.

[Batstone et al., 2002] Batstone, D., Keller, J., Angelidaki, I., Kalyuzhnyi, S., Pavlostathis, S., Rozzi, A., Sanders, W., Siegrist, H., and Vavilin, V. (2002). *Anaerobic Digestion Model No.1 (ADM1)*. IWA Publishing.

[Bear, 1972] Bear, J. (1972). *Dynamics of Fluid in Porous Media*. Dover Publications, INC. New York.

[Bear and Bachmat, 1990] Bear, J. and Bachmat, Y. (1990). *Introduction to Modeling of Transport Phenomena in Porous Media*. Kluwer Academic Publishers, Dordrecht, Boston, London.

[Bey, 1997] Bey, J. (1997). *Finite-Volumen- und Mehrgitterverfahren für elliptische Randwertprobleme*. PhD thesis, Eberhard-Karls-Universität Tübingen.

[Boubaker and Ridha, 2008] Boubaker, F. and Ridha, B. C. (2008). Modelling of the mesophilic anaerobic co-digestion of olive mill wastewater with olive mill solid waste using anaerobic digestion model No. 1 (ADM1). *Bioresource Technology*, 99(14):6565 – 6577.

[Chynoweth et al., 2001] Chynoweth, D. P., Owens, J. M., and Legrand, R. (2001). Renewable methane from anaerobic digestion of biomass. *Renewable Energy*, 22(1-3):1 – 8.

[Deuflhard and Bornemann, 2002] Deuflhard, P. and Bornemann, F. (2002). *Numerische Mathematik: Numerische Mathematik 2: II.* de Gruyter Lehrbuch.

[Deuflhard and Hohmann, 1991] Deuflhard, P. and Hohmann, A. (1991). *Numerische Mathematik: Eine algorithmisch orientierte Einführung*. de Gruyter Lehrbuch.

[Dziuk, 2010] Dziuk, G. (2010). *Theorie und Numerik partieller Differentialgleichungen*. DeGruyter.

[Gilbarg and Trudinger, 1998] Gilbarg, D. and Trudinger, N. S. (1998). *Elliptic Partial Differential Equations of Second Order*. Springer.

[Grillo et al., 2010] Grillo, A., Logashenko, D., Stichel, S., and Wittum, G. (2010). Simulation of density-driven flow in fractured porous media. *Adv. Water Resour.*, 33(12):1494–1507.

[Hassanizadeh and Gray, 1979a] Hassanizadeh, S. and Gray, W. (1979a). General conservation equations for multiphase systems: 1. Averaging procedure. *Adv Water Resour*, 2(3):131–144.

[Hassanizadeh and Gray, 1979b] Hassanizadeh, S. and Gray, W. (1979b). General conservation equations for multiphase systems: 2. Mass, momenta, energy and entropy equations. *Adv Water Resour*, 2(4):191–203.

[Hassanizadeh and Gray, 1980] Hassanizadeh, S. and Gray, W. (1980). General conservation equations for multiphase systems: 3. Constitutive theory for porous media flow. *Adv Water Resour*, 3(1):25–40.

[Lemke et al., 2012] Lemke, B., Logaschenko, D., Muha, I., Nägel, A., Paech, B., Wehner, C., and Wittum, G. (2012). The software system ug3 for the simulation of pde-based models. *Computing and Visualization in Science*. In preparation.

[Muha, 2012] Muha, I. (2012). *Modellierung und Simulation eines Biogasreaktors*. PhD thesis, Goethe-Universität Frankfurt am Main.

[Muha et al., 2012] Muha, I., Grillo, A., Heisig, M., Schönberg, M., Linke, B., and Wittum, G. (2012). Mathematical modeling of process liquid flow and acetoclastic methanogenesis under mesophilic conditions in a two-phase biogas reactor. *Bioresource Technology*, 106:1–9. DOI: 10.1016/j.biortech.2011.11.087.

[Muha et al., 2013] Muha, I., Zielonka, S., Lemmer, A., Schönberg, M., Linke, B., Grillo, A., and Wittum, G. (2013). Do two-phase biogas plants separate anaerobic digestion phases? a mathematical model for the distribution of anaerobic digestion phases among reactor stages. *Bioresource Technology*, 132(0):414 – 418.

[Nägel et al., 2008] Nägel, A., Hansen, S., Neumann, D., Lehr, C.-M., Schaefer, U. F., Wittum, G., and Heisig, M. (2008). In-silico model of skin penetration based on experimentally determined input parameters. Part II: Mathematical modelling of in-vitro diffusion experiments. Identification of critical input parameters. *Eur J Pharm Biopharm*, 68(2):368–379.

[Ortega and Rheinboldt, 1987] Ortega, J. M. and Rheinboldt, W. C. (1987). *Iterative Solution of Nonlinear Equations in Several Variables (Classics in Applied Mathematics)*. Society for Industrial Mathematics.

[Reiter and Wittum, 2014] Reiter, S. and Wittum, G. (2014). Promesh - a flexible interactive meshing software for unstructured hybrid grids in 1, 2 and 3 dimensions. *in preparation*.

[Schönberg and Linke, 2009] Schönberg, M. and Linke, B. (2009). Vergärung von Roggen-Ganzpflanzensilage in einer zweiphasigen Prozessführung - Anaerobic digestion of rye silage in two-phase process. *Bornimer Agrartechnische Berichte*, 68:114–125.

[Schönberg and Linke, 2012] Schönberg, M. and Linke, B. (2012). The influence of the temperature regime on the formation of methane in a two-phase anaerobic digestion process. *Engineering in Life Sciences*, 12(3):279–286.

[Thamsiriroj and Murphy, 2011] Thamsiriroj, T. and Murphy, J. (2011). Modelling mono-digestion of grass silage in a 2-stage CSTR anaerobic digester using ADM1. *Bioresource Technology*, 102(2):948 – 959.

[Vogel et al., 2012] Vogel, A., Reiter, S., Rupp, M., Nägel, A., and Wittum, G. (2012). Ug 4 - a novel flexible software system for simulating pde based models on high performance computers. *CVS*. In preparation.

[Wett et al., 2007] Wett, B., Schoen, M., Phothilangka, P., Wackerle, F., and Insam, H. (2007). Model-based design of an agricultural biogas plant: application of anaerobic digestion model no. 1 for an improved four chamber scheme. *Wat. Sci. Tech.*, 55(10):21–28.

[Wichern et al., 2009] Wichern, M., Gehring, T., Fischer, K., Andrade, D., Lübken, M., Koch, K., Gronauer, A., and Horn, H. (2009). Monofermentation of grass silage under mesophilic conditions: Measurements and mathematical modeling with ADM1. *Bioresource Technology*, 100:1675–1681.

[Zielonka et al., 2010] Zielonka, S., Lemmer, A., Oechsner, H., and Jungbluth, T. (2010). Energy balance of a two-phase anaerobic digestion process for energy crops. *Engineering in Life Sciences*, 10(6):515–519.

Reviewed by:

- Prof. Dr. Günther Busch, BTU Cottbus-Senftenberg, Germany
- Prof. Dr. Bernd Linke, Leibniz Institute for Agricultural Engineering Potsdam-Bornim, Germany
- Prof. Dr. Alfio Grillo, Politecnico di Torino, Italy

In: Methanogenesis
Editor: Gagik Badalians Gholikandi
ISBN: 978-1-63321-567-2

Chapter 5

BIOGAS PRODUCTION WITH EMPHASIS ON METHANOGENESIS PROCESS: AN INVESTIGATIVE VIEW TO APPLICATIONS OF ARTIFICIAL NEURAL NETWORKS MODELING

Mohammad Azimipour, Gagik Badalians Gholikandi [*]
and Hamidreza Masihi
Faculty of Water Engineering and Environment,
Shahid Beheshti University, A.C., Tehran, Iran

ABSTRACT

The anaerobic digestion (AD) process, a chain of interconnected biological reactions, can be viewed as the biodegradation of organic matter. This process consists of two main biochemical successive process-stages and is performed by various groups of bacteria in an oxygen-free environment. Consequently, biogas (i.e., methane and carbon dioxide) production occurs permanently. For digestion to occur properly, methanogenesis process is necessary to convert the organic acids to biogas. The anaerobic process is very sensitive to environmental and operation conditions of temperature, pH, nutrients concentration,

* Corresponding author: Hakimieh- Tehranpars, Shahid Abbaspour Blvd., PO Code 16589-53571, PO Box 16765-1719, Tehran, Iran. Email: g.badalians@yahoo.com.

hydraulic retention time, and concentration of volatile fatty acids. Owing to complexity of the anaerobic digestion process, there has been a growing demand to simulation and monitoring by means of different modeling approaches. Artificial neural network (ANN) is one of these methods that have been employed to simulate engineered environments performance in different operation conditions. In this chapter researches results regarding anaerobic digestion process modeling have been investigated comprehensively. Furthermore, by using data which are obtained from another study, two various techniques of artificial neural networks modeling, i.e., Multi-Layer Perceptron (MLP) and Self Organizing Feature Maps (SOFM) were implemented and different possible applications were proposed. The results of this investigation represent the capability of the artificial neural networks approach to predict and optimize the anaerobic digestion process performance efficiently.

Keywords: Anaerobic digestion (AD), Methanogenesis, Artificial neural network, Multi-Layer Perceptron (MLP), Self Organizing Feature Maps (SOFM)

1. INTRODUCTION

The anaerobic digestion process is a chain of interconnected biological reactions. Thus, the available organic matter converts into methane, carbon dioxide and anaerobic biomass in an oxygen-free environment (Bravo et al., 2011). This process consists of four successive steps; each step is characterized by the activity of a certain group of bacteria and is sensitive to specified environmental conditions (Levstek and Lakota 2010). In many cases, the biochemical reactions during the anaerobic digestion process performance has been classified according to two main groups, i.e., heterogeneous (hydrolysis) and homogeneous (acidogenesis, acetogenesis and methanogenesis) (Mara and Horan 2003; Pipoli 2005; Gerardi 2006; Apples et al., 2008; GTZ 2009; Maghanaki et al., 2013; Yu et al., 2013).

Processing a successful anaerobic digestion requires the presence of a large number of methane-forming bacteria and the wide diversity of the related microorganisms (Gerardi 2003). The three most important groups of the methane-forming bacteria are hydrogenotrophic, acetotrophic, and methylotrophic methanogens (Badalians Gholikandi 2009). About 70% (two-thirds) of the methane yield results from acetate conversion by acetotrophic methanogens and the rest is the result of carbon dioxide reduction by means of hydrogen (Badalians Gholikandi 2009; Mackie and Bryant 1981).

The acid-splitting methane-forming bacteria have the slowest growth rate. Therefore, they attract attention as the process rate limiting agent (Parkin and Owen 1986; Moletta 1986). Moreover, methane bacteria have a slower growth rate than the hydrolytic bacteria due to lower energy generation.

In an anaerobic environment, various parameters affect the reaction rates of different steps such as follows (Apples et al., 2008):

Temperature

The anaerobic digestion process performance in municipal wastewater treatment plants occurs in the temperature range of 25 to 40°C (mesophilic condition); optimum condition equals 35°C (Bitton 2005). The temperature range of 50 to 65°C is determined for the thermophilic digestion's condition which makes higher loading rates feasible (Bitton 2005). Due to slower growth of methane-forming bacteria, they are very sensitive to small changes in the temperature (Badalians Gholikandi 2009).

pH

Different optimum pH ranges have been determined according to related groups of microorganisms (Apples et al., 2008). Methane bacteria have an optimum pH range of 6.5 to 7.2 (Boe 2006; Turovskiy and Mathai 2006). Thus, the related range to the fermentative bacteria equals 4 to 8.5 which indicates that these species aren't very sensitive (Hwang 2004). Acetic and butyric acids are the main products of the first process stage at a low pH. However, acetic and propionic acids are the main ones at a high pH, i.e., pH≥8 (Boe 2006).

Hydraulic Retention Time (HRT)

The hydraulic retention time depends considerably on wastewater characteristics and environmental conditions (Bitton 2005). The attached growth environment requires hydraulic retention time equals 1 to10 days; the suspended growth medium 10 to 60 days (Polprasert 1989). Due to the increasing volatile fatty acids (VFA) concentration, retention times less than 5 days are not sufficient; this is because of the washout of methane bacteria (Apples et al., 2008). If the sludge retention time equals 5 to 8 days, VFA concentration is still high which causes an incomplete breakdown of the existing compounds, especially lipids (Apples et al., 2008). Therefore, a sludge retention time (SRT) equals 8 to10 days is required to reach the necessary stable state (Figure 1).

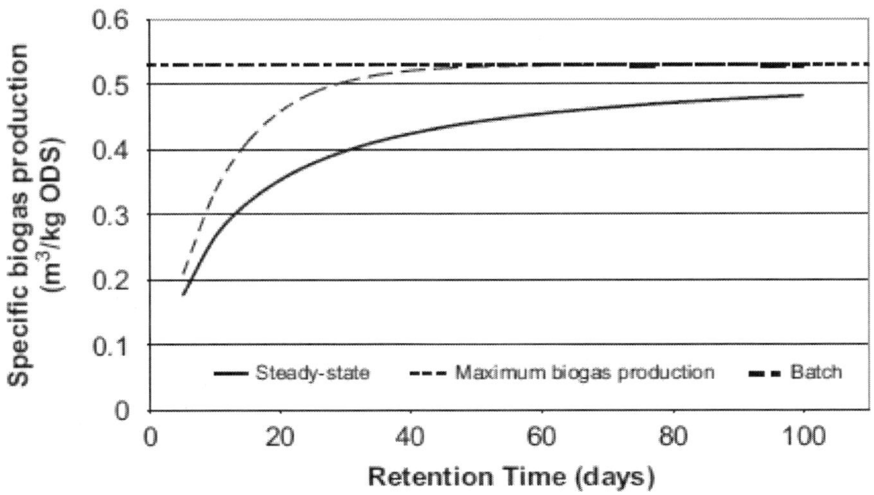

Figure 1. Biogas production vs. Sludge retention time (SRT) (Apples et al., 2008).

Chemical Composition of Wastewater

Finding a balance between nutrients contained in wastewater, i.e., phosphorous, nitrogen and sulfur is essentially for an adequate anaerobic digestion process performance (Bitton 2005). The appropriate ratio of C: N: P is 700:5:1 accordingly (Lettinga 1995; Sahm 1984). Moreover, the ratio of C: N equals 25-30:1 means that the optimal gas production is assured (Polprasert 1989; El-Hinnawi and Biswas 1981). It should be considered that methanogens require sulfide as well. However, if its concentration exceeds 150-200 mg/L, it will have toxic effect on these species (Speece 1983).

Volatile Fatty Acid (VFA)

High concentrations of the VFA cause inhibition or failure of the anaerobic process. It has been shown that the VFA concentrations more than 4 g/L cause inhibited glucose fermentation (Ward et al., 2008). Moreover, a digestion process failure happens if the propionic acid concentration exceeds 3000 mg/L (Boone and Xun 1987). However, it has been indicated that this acid is an effect rather than a cause of inhibition (Pullammanappallil et al., 1998). Toxic effect of the propionic acid on acid-forming bacteria is considerable (Bitton 2005). It should be considered that excessive fatty acids are toxic only in their non-ionized form which is pH-dependent (Ward et al., 2008). If the pH range is maintained close to neutrality (Bitton 2005) or less than 7 (Mata-Alvarez 2002) volatile fatty acids are toxic.

Due to the complexity of the anaerobic digestion process, an application of monitoring and a control system is necessary to avoid process destabilization and to keep a stable state (Molina et al., 2009). Furthermore, in some cases, operators reject valuable waste that can produce high methane volume because they are worried that serious disturbance of the biological system can be occurred (Holubar et al., 2002). Thus, there is a need of an on-line model for predicting the behavior of the system. Therefore, an alternative to optimize the anaerobic digestion process design and operation is to employ efficient modeling tools. Many reasons motivate us to apply the modeling as follows:

1) Sensitivity of the anaerobic fermentation process to operating conditions such as pH, temperature, overloads and etc. (Hernandez 2013).
2) The parameters cannot be determined precisely because of specific features of living organisms (Hill and Barth 1977; Dochain and Vanrolleghem 2001) that makes the system unstable (Yordanova 2004).
3) Prototype requirements of systems are expensive and the related studies are time-consuming (Yu et al., 2013).
4) Expensive biomass sensors which are not necessarily designed from an automatic control perspective (Hernandez 2013).
5) Reactors are designed basically for relatively stable operating conditions, but because of fluctuations, these nominal conditions are seldom met in practice.

6) The AD process remains a black box (Dochain 1995; Steyer et al., 1995) due to complex mechanisms ruling this process (Al Seadi 2009).
7) Inadequate understanding of biochemical and physical processes involved in anaerobic treatment often results in low CH_4 production, accumulation of volatile fatty acids (VFAs) in the digesters (Güclü et al., 2011) and a drop in pH (Wilcox et al., 1995).
8) It is difficult to construct a model that completely describes the process due to the interaction of microorganisms in the digestion process (Wilcox et al., 1995).
9) Differential equations require significant subjective assumptions, at least on the selection of the meaningful parameters for modeling (Kana et al., 2012).

Because of the aforementioned issues, there has been a growing demand for improving the performance of AD process by applying an efficient knowledge extraction and diagnosis technique. Reliable mathematical models can be of great assistance to facilitate experiments, improvements of the process and offer great possibility to control it. Therefore, several attempts have been done in this area (Holubar et al., 2002). But this kind of modeling has some drawbacks (Güclü et al., 2011):

1) Large number of parameters that should be estimated for model implementation.
2) The estimated parameters are case specific and difficult to adopt for system modification.
3) Although it is essential to define the biodegradable fraction of the influent for accurate digester simulation, the organic composition and the amount of influent stream is impossible to quantify.

One of the most popular mathematical modeling methods of anaerobic digestion process is ADM1 (Anaerobic Digestion Model NO.1). This platform was created in 1988 by a task group in the International Water Association (IWA). Several researchers have used this model and have pointed out some weaknesses in ADM1 such as limitations in simulating complex processes and requiring a detailed characterization of the substrate which might not be possible to do on a regular basis (Blumensaat and Keller 2005; Derbal et al., 2009; Lubken et al., 2007; Daels et al., 2009; Kleerebezem and Van Loosdrecht 2006).

Due to the drawbacks of different modeling methods, another approach is required for better operation of biogas digesters to maximize methane production. Therefore, modeling studies have partly been shifted to stochastic approaches (Kanat and Saral 2009). Artificial Neural Network (ANN) has received considerable attention for this goal, because:

1) They have attributes that make them potentially successful in simulation and predictions (Abu Qdais et al., 2010).
2) They are capable of learning complex highly nonlinear relations (Abu Qdais et al., 2010) and their tolerance for failure is significant (Güclü et al., 2011).
3) They are capable of associations from a large body of data due to their intrinsic nonlinearity, adaptability, noise immunity, generalization ability and robustness (Abu Qdais et al., 2010).
4) They can capture and represent complex input/output relationships (Holubar et al., 2002).
5) They have distinctive advantages over some other nonlinear estimation methods (Holubar et al., 2002).
6) They don't require prior knowledge about the structure and relationships (Strik et al., 2005).
7) They can combine required sensors which detect short-time scale events (Wilcox, et al., 1995).
8) They can recognize catastrophic events rapidly (Wilcox, et al., 1995).
9) They can be fitted to any kind of data set and did not require model assumptions (Savegnago et al., 2011). They also can handle multivariable problems (Dong and Zhao, 2013).
10) They have found their role into many applications and the ability to perform nonlinear pattern recognition, classification and function approximation (Parthiban et al., 2007).

Following are the examples of some recent works done in the area of environmental engineering which indicate the high capability of this approach. Djebbar and Narbaitz used the neural network model for the analysis of the mass transfer coefficient in air stripping packed towers which is applied for removing volatile organic compounds (Djebbar and Narbaitz 2002). Hamed et al. applied neural networks in wastewater treatment plants predictions (Hamed et al., 2004). Hong et al. utilized KSOFM to diagnose the inter-relationship of the process variables in a real activated sludge (Hong et al., 2003). Tobiszewski et al. used neural networks to assess the efficiency of chlorinated solvent removal from petrochemical wastewater in a refinery wastewater treatment plant (Tobiszewski et al., 2012). Grieu et al. presented integration of neural networks in a geographical information system for developing rain flow predictive models (Grieu et al., 2006). Pendashteh et al. applied ANN for modeling of membrane bioreactor treating hypersaline oily wastewater (Pendashteh et al., 2011). Khataee and Kasiri used the artificial neural network for modeling of contaminated water treatment processes by homogeneous and heterogeneous nanocatalysis (Khataee and Kasiri 2011). Badalians Gholikandi and Khsoravi used the artificial neural networks modeling in upgrading of submerged membrane bioreactor operation with regard to soluble microbial products and in optimization of critical flux (Badalians Gholikandi and Khosravi 2012). Badalians Gholikandi et al. predicted the biological UCT process efficiency by using artificial neural network (Badalians Gholikandi et al., 2011a). Badalians Gholikandi et al. applied the artificial neural networks model in prediction of coagulation/flocculation process by PAC in water treatment plant (Badalians Gholikandi et al., 2011a). Badalians Gholikandi et al. predicted pressure filters performance and determination of optimum turbidity for coli-form and total bacteria removal by using ANNs (Badalians Gholikandi et al., 2011b). Badalians Gholikandi et al. applied both MLP and KSOFM neural networks for Optimizing stabilization of waste-activated sludge using Fered-Fenton process (Badalians Gholikandi et al., 2014a). Badalians Gholikandi et al. utilized MLP neural networks for optimization of the anaerobic baffled reactor (ABR) in municipal wastewater treatment (Badalians Gholikandi et al., 2014b).

2. Artificial Neural Networks

The artificial neural networks are robust mathematical tools which are inspired from the biology of the human brain. A neural network is a system of simple processing elements called neurons (Strik et al., 2005) which are arranged in layers (Holubar et al., 2002). Each neuron is connected to its neighbors with varying coefficients called weights that represent the relative influence of different neuron inputs to other neurons (Hamed et al., 2004). A prescribed set of well-defined rules for the solution of a learning problem is called learning algorithm (Haykin 1994). These algorithms differ from each other in the way in which the adjustment to a weight of a neuron is formulated.

The artificial neural networks can be classified in many different ways based on various characteristics. The main feature of this classification is learning algorithm; we have supervised and unsupervised manner accordingly.

The most widely used supervised learning algorithm which is considered the workhorse of ANNs (Rumelhart et al., 1986) is backpropagation. The specific characteristic of this kind of neural network is that they require a teacher that may be a training set of data. The aim of this learning is to minimize the error between network outputs and real targets by modifying the weights. The most popular network that uses this kind of learning algorithm is Multi-Layer Perceptron (MLP). Generally, this network consists of these layers: input layer, hidden layer and output layer. Previous researches have shown that one hidden layer is sufficient for environmental issues (Rahmanian et al., 2011). Moreover, due to the universal approximation theory, a network with a single hidden layer with a sufficiently large number of neurons can be used for any input-output structure interpretation (Salari et al., 2005). In an MLP network, firstly inputs are multiplied by weights. Then they are passed through the transfer function and finally outputs are produced. This process is accomplished in hidden layer neurons. After that the hidden neurons outputs are fed to the output layer and the same process will be applied. The cascade of layers can be thought of as a black box in which the knowledge of an ANN is stored in its weights (Holubar, et al., 2002). One of the most popular training algorithms is Levenberg-Marquardt (L-M). This kind of algorithm is a combination of the steepest descent method and the Gauss–Newton algorithm which inherits the speed advantage of the Gauss–Newton algorithm and the stability of the steepest descent method. The L-M method is 10 to 100 times faster than the usual gradient descent back propagation method (Pendashteh et al., 2011).

While the number of input and output neurons depends on the nature of the problem (Salari et al., 2005), the numbers of hidden layer neurons will be determined in a trial-and-error approach which includes of constructing different networks by varying the number of neurons and the activation functions. Different functions have been used in previous studies, but three main ones are tan-sigmoid, log-sigmoid and linear. The sigmoid functions especially tan-sigmoid are needed to introduce non-linearity into the network (Hamed et al., 2004). Due to suitability of linear functions for continuous valued targets (Sarle 2002), they have been selected for output neuron.

For evaluating the performance of the models and finding the best fitting option, various indices are proposed such as Mean Square Error (MSE), Root Mean Square Error (RMSE), Mean Absolute Error (MAE), Mean Absolute

Percentage Error (MAPE), R (Regression Coefficient) and R^2 (Correlation Coefficient). In this survey, MSE, R and R^2 were chosen because of the reviewed literature and are shown in equations (1), (2) and (3).

$$MSE = \frac{1}{N}\sum_{i=1}^{N}(T_i - O_i)^2 \tag{1}$$

$$R = \frac{\sum_{i}^{n}(T_i - \overline{T_i})(O_i - \overline{O_i})}{\sqrt{\sum_{i}^{n}(T_i - \overline{T_i})^2 \sum_{i}(O_i - \overline{O_i})^2}} \tag{2}$$

$$R^2 = 1 - \left[\frac{\sum_{i=1}^{n}(T_i - O_i)^2}{\sum_{i=1}^{n}T_i^2}\right] \tag{3}$$

T and O indicate experimental and network outputs, $\overline{T}$ and $\overline{O}$ are the average of mentioned data, respectively. And N is the total number of data.

The second category of ANNs based on the learning algorithm is unsupervised networks that use no external teacher and are based upon local information (Hamed et al., 2004). Instead, there is an internal monitoring of performance. This kind of network looks for regularities or trends in the input signals. To perform unsupervised learning, a competitive learning rule may be used in which neurons compete with each other. In its simplest form, the network operates in accordance with a "winner-takes-all" strategy (Haykin 1994) in which the winning neuron or the best matching unit (BMU) will be chosen. Such competition can be induced by having lateral inhibition connections (negative feedback paths) between the neurons. The result is that the neurons are forced to organize themselves. Because of these reasons, such a network is called a Self Organizing Map (SOM).

The most widely used unsupervised neural network is Self Organizing Feature Maps (SOFM) which is used to reduce the amount of data by clustering and to project the data nonlinearity onto a lower dimensional space (Hong et al., 2003). This network consists of two layers: input layer and the competitive layer in which neurons are fully connected. The neurons of this kind of neural network are located on a regular low-dimensional grid. The lattice of the grid can be either hexagonal or rectangular (Vesanto 1999) that the former is used in this chapter. Figure 2 shows two different kinds of the lattice.

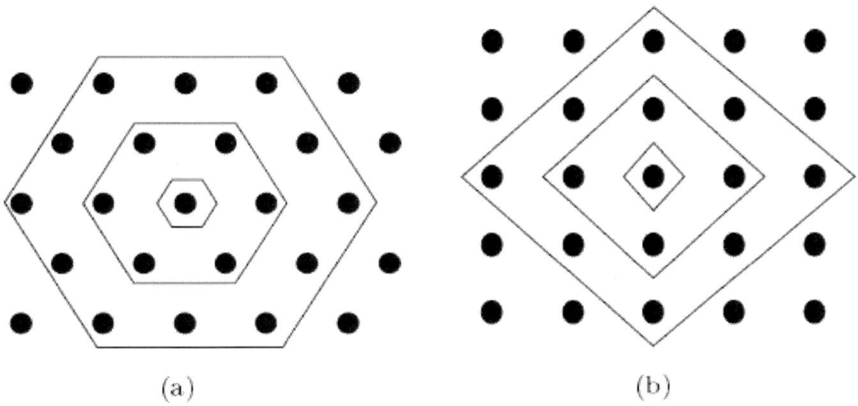

Figure 2. (a) Hexagonal lattice, (b) rectangular lattice (Vesanto 1999).

Each neuron k is represented by an n-dimensional prototype vectors (weights) $m_k = [m_{k1},\ldots,m_{kn}]^T$, where n is the dimension of the input space. For training this kind of neural network there are two algorithms: (1) sequential training in which samples are presented to the map one at a time, and the algorithm gradually moves the weight vectors towards them and (2) batch training in which the data set is presented to the SOM as a whole, and the new weight vectors are weighted averages of the data vectors (Vesanto 1999). When specified errors are attained, the training process will be stopped. The following two error criteria are normally used: the quantization error and the topographical error, as defined by Eq. 4 and 5, respectively (Garcia and Gonzalez 2004).

$$q_e = \frac{1}{N}\sum_{i=1}^{N} \| x_i - m_c \| \quad (4)$$

$$t_e = \frac{1}{N}\sum_{i=1}^{N} u(x_i) \quad (5)$$

q_e= quantization error
N = number of samples
x_i = i-th data sample or vector;
m_c = prototype vector of the best matching unit for x_i;
||…|| denotes distance;
t_e = topographical error;
u = binary integer, which is equal to 1 if the first and second BMU for the argument of u are not adjacent units on the map, otherwise, i = 0.

For finding the side length of any map, a heuristic formula is applied which is proposed in Eq. 6 (Garcia and Gonzalez 2004).

$$M = 5\sqrt{N} \quad (6)$$

Where M is the number of map units and N is the number of training data samples. In this paper 18 samples were used. Therefore, there should be at least 20 map units.

3. Applications of Artificial Neural Networks in Biogas Production

In recent years ANNs have been increasing applied for optimizing biogas production in the AD process. Selecting parameters for inputs and outputs of the network, training algorithm, transfer functions, structure of networks and the results of modeling are the main factors that should be considered for better simulation. In this section we gathered aforementioned factors from different studies that have been done according to the biogas production, especially methane production; finally, a short review will be proposed regarding these studies in the Table 2.

Abu Qdais et al. used the ANN for simulating and optimizing of the biogas production process. The parameters considered in this study were temperature (T), total solids (TS), total volatile solids (TVS) and pH. Back propagation algorithm was used for training the MLP network which has two hidden layers and each one has 25 neurons. Therefore, the structure of this network was 4-25-25-1. The output of this network was methane production. The values of MSE and R^2 for this network were 6×10^{-5} and 0.87, respectively. This survey has shown that predicted values of the methane fit well with the measured values (Abu Qdais et al., 2010).

Holubar et al. used three neural networks to control methane production in an anaerobic digestion. The parameters used for modeling are: gas production (GP), gas composition (GC), volatile fatty acids (VFA), pH, organic loading rate (Br), acetic acid (AA) and propionic acid (PA). The selected model for the first simulation had the structure of 3-3-1. The input variables were Br_{t-1}, Br_{t-2}, pH_{t-1} and the output variable was pH_t. The regression coefficient for this model was 0.82. The structure of selected model for the second simulation was 9-3-3 in which inputs were Br_t, Br_{t-1}, Br_{t-2}, pH_{t-1}, GP_{t-1}, GC_{t-1}, VFA_{t-1}, AA_{t-1}, PA_{t-1} and outputs were total of VFA (VFA_{tot}), AA_t and PA_t. The regression coefficient for VFA_{tot}, AA_t and PA_t were 0.86, 0.84 and 0.85, respectively. The structure of the third model was 9-3-2. For this modeling, inputs were as same as the second model, but the outputs were GP_t and GC_t which had a regression coefficient of 0.9 and 0.8, respectively. In this study, t and t-1 indicated the measure at time t and one time step backwards; t+1, t+2 and t+3 indicated the prediction one step forward, two steps forward and three steps forward, respectively. This research has shown that the anaerobic digestion of surplus sludge can be effectively modeled by means of a hierarchical system of neural networks. Therefore, a prediction of gas production and composition can be made several time-steps in advance to predict and avoid shock loads (Holubar et al., 2002).

In another research, Strick et al. utilized ANN to predict trace compounds like hydrogen sulfide (H_2S) and ammonia ($N\text{-}NH_3$). The first model used for H_2S prediction had taken sulfate loading rate ($[SO_4^{2-}]_{t-1}$, $[SO_4^{2-}]_t$, $[SO_4^{2-}]_{t+1}$), organic loading rate ($[COD]_t$, $[COD]_{t+1}$) and hydrogen sulfide ($[H_2S]_t$, $[H_2S]_{t-1}$) as the inputs and H_2S ($[H_2S]_{t+1}$) as the output. The structure of this model was 7-5-1. Tansig (tan-sigmoid) and purelin (linear) were used as the transfer functions for hidden and output layer, respectively. Traingdm (batch gradient descent with momentum algorithm) was utilized as the training algorithm. The indices used for evaluating the performance of this model were R^2 and RMSE by which for training data were 0.82 and 1062 and for validation data were 0.91 and 416, respectively (Strik et al., 2005). The second model used for ammonia prediction utilized total nitrogen ($[N]_{t+1}$), organic loading rate ($[COD]_{t+1}$), $[\text{biogas productivity}]_t$, $[pH]_t$, $[pH]_{t-1}$, $[pH]_{t-2}$, ammonia in biogas ($[NH_3]_t$) and ammonia in the reactor ($[N\text{-}NH_3]_t$) as the inputs and ammonia in biogas ($[N\text{-}NH_3]_{t+1}$) as the only output. The structure of the best fitted model was 8-7-1. Activation functions and training algorithm was as same as the first model. R^2 and RMSE for training data were 0.82 and 8 and for validation data were 0.83 and 5.1, respectively. This paper confirmed that trace compounds in biogas from an anaerobic digestion were modeled successfully using Matlab neural network toolbox representing that developed neural networks are suitable in predictive control tools and an operator can take the right precautions in a few minutes (Strik et al., 2005).

Hawkes et al. used neural networks for monitoring and controlling an AD process by utilizing data from an on-line bicarbonate alkalinity (BA) sensor. The structure of this network was 8-8-1 which indicated eights inputs containing the last eight BA data points and the output was likelihood of the process being out of control (Wilcox et al., 1995).

Ozkaya et al. presented a neural network model for predicting the methane fraction in landfill gas originating from field-scale landfill bioreactors. Two models were proposed. For the first model, input variables were pH, alkalinity, conductivity, cl^-, chemical oxygen demand (COD), sulfate (SO_4^{2-}) of leachate, waste temperature (T) and refuse age and the output was the methane fraction with leachate recirculation. The structure of the selected model was 8-12-1. The regression coefficient was 0.951 and MSE was 0.0026. For the second model input variables were as same as the first model, but the output was the methane fraction without leachate recirculation. The structure of the selected model was 8-15-1. For both models 10 BP training algorithms were used, but L-M was chosen due to the better performance. Moreover, tan-sigmoid and linear activation functions were used for the hidden and output layer, respectively. This study has shown the high capability of ANNs in the modeling of the methane production by considering environmental factors such as water addition (recirculation), waste temperature and leachate components. The authors also suggested the development of ANNs in terms of (i) predicting hourly methane production based on leachate parameters, (ii) control strategies can be extended with the objective of increasing methane production, (iii) providing optimization of the conversion to energy of methane gas in landfills, (iv) determining the optimum construction time of a waste energy plant at landfill sites, and (v) determining leachate recirculation strategies and strategies for reducing of greenhouse gases (Ozkaya et al., 2007).

Kianmehr et al. used the ANN to predict sludge biodegradability which was measured by biochemical methanogenic potential (BMP[1]) test. For finding the best fitted model, different combinations of studied parameters: Sonication Intensity (SI), sludge solids residence times (SRT), volatile dissolved solids (VDS), filtered chemical oxygen demand (FCOD) and filtered total Kjeldhal Nitrogen (FTKN) were tested which is summarized in Table 1. The regression coefficients of generated networks were in range of 0.81-0.98 and the L-M was chosen as the training algorithm (Kianmehr et al., 2014).

Table 1. Combinations of the studied parameters (Kianmehr et al., 2014)

Combination	Description
0	All Parameters
1	SRT+SI+VDS+FCOD (FTKN Removed)
2	SRT+SI+VDS+FTKN (FCOD Removed)
3	SRT+SI+FCOD+FTKN (VDS Removed)
4	SRT+SI+VDS (FTKN and FCOD Removed)
5	SRT+SI+FCOD (FTKN and VDS Removed)
6	SI+VDS+FCOD (FTKN and SRT Removed)

This survey indicated the capability of ANNs in predicting biogas generation profiles of WAS and the following conclusions could be drawn from the research preliminary-results (Kianmehr et al., 2014):

- The best accuracy is achieved when SRT, SI, VDS and FCOD parameters are included as inputs for the neural network.
- The VDS parameter is the major predictor followed by the FCOD parameter when SRT and SI values are known.
- The FTKN parameter has little contribution in enhancing the neural network prediction accuracy.
- Distorted prediction results could be obtained from a neural network having excess input parameters.

[1] BMP assay provides information on the potential extent and rate of methane production available from a specific feedstock (Ward et al., 2008).

- The SRT parameter is a major parameter that cannot be excluded, even if both VDS and FCOD parameters were included.

Güclü et al. implemented a model to simulate the dynamic behavior of a large-scale anaerobic sewage sludge digestion system. Two basic models were trained to predict volatile solids (VS) and methane production. Each model consists of four various models. The differences of these models were in the structure of networks (one and two hidden layers) and training algorithm in which Gradient descent (GD) with adaptive learning rate and Levenberg-Marquardt were utilized. For models used for volatile solids prediction, inputs were sludge flow rate (Q), temperature (T), pH, volatile fatty acids of influent (VFA) and alkalinity and the output was volatile solids of effluent (VS). And for methane production models, besides first model inputs, volatile solid of influent was considered as the sixth input. The range of regression coefficient of models applied for the VS prediction was 0.84-0.9 for training data and 0.83-0.89 for testing data. MSE was 1.12-1.93 for training data and 1.2-2 for testing data. The range of regression coefficient of models developed for methane production was 0.75-0.88 for training data and 0.63-0.71 for testing data. MSE was 1.17-1.5 for training data and 1.2-2 for testing data. This study has shown that ANNs can be capable of dynamically predicting the VS and CH_4 production rates of the real system and therefore, these models can be applied to monitor changes in the anaerobic digestion. Furthermore, only relatively simple monitoring parameters are needed to build a model of the high complex anaerobic digestion process (Güclü et al., 2011).

Kanat and Saral used MLP as the modeling tool to estimate the biogas production rate in a thermophilic upflow anaerobic sludge blanket digester. The inputs of this model were organic loading rate (OLR), total volatile fatty acids (VFA), alkalinity of influent and effluent, pH of influent and effluent and temperature (T). This study has shown that ANN models gave encouraging estimation results for the online control of thermophilic reactors in both conditions of steady state and abnormal conditions (Kanat and Saral 2009).

Dong and Zhao investigated the suitability and accuracy of modeling the rumen methane production of mixed rations for cattle by using ANN. Three models were developed to predict methane, carbon dioxide and total gas production. The inputs of these models were sugars (CA), starch and pectin (CB_1) and available cell wall (CB_2). All models had used L-M as training algorithm. For the first model which has methane production as the output, values of R^2 and RMSE (%) were 0.95 and 3.89%, respectively. The structure of the best fitted model was 3-8-1 (Dong and Zhao, 2013). The second model by the structure of 3-11-1 had found R^2 and RMSE (%), 0.97 and 2.95%, respectively. For third model for predicting the total gas production, values of R^2 and RMSE (%) were 0.92 and 4.23%. This paper has indicated that BP models have shown similar accuracy with the multiple regression model for predicting the CH_4 production and better accuracy for predicting the CO_2 and the total gas production than the multiple regression models (Dong and Zhao 2013).

Parthiban et al. utilized MLP neural network in which inputs were organic loading rate (OLR) and influent pH and outputs were effluent chemical oxygen demand (COD), biogas, volatile fatty acids (VFA), alkalinity and effluent pH. This model had two hidden layers and the value of MSE was 0.0146. The utilized model in this paper has identified the nonlinear relationship between the various experimental parameters (Parthiban et al., 2007).

Kana et al. studied the optimization of the biogas production by using ANN on the mixed substrates of saw dust (SD), cow dung (CD), banana stem (BS), rice bran (RB) and paper waste (PW). While mentioned parameters were the inputs, biogas performance index of the digester was the output. Sigmoid transfer functions and the L-M algorithm were used for this model that had the structure of 5-2-1 and the MSE less than 0.01. In this paper, artificial intelligence based tools can significantly reduce the process development tool (Kana et al., 2012).

Badalians Gholikandi et al. worked on optimization of anaerobic baffled reactor (ABR) using ANN in the municipal wastewater treatment in which COD, TSS, organic loading rate (OLR), temperature (T), and volatile fatty acids (VFA) were taken as the inputs and COD and VFA of output were regarded as the network outputs. Tan-sigmoid and traingdm (train gradient descent method) were used for this network which has a structure of 6-12-2-2. The correlation coefficient (R) of the training and testing were 0.997 and 0.974, respectively. The conclusion of this work based on ANN modeling is that the optimized ABR should be designed with a six-compartment and operated at 10-12 (hr) HRT (Badalians Gholikandi et al., 2014a).

Elnekave et al. used ANNs for predicting COD and TSS concentrations in effluent of upflow anaerobic sludge blanket (UASB) reactor treating citrus juice wastewater. The inputs of networks were flow, volumetric load, COD (influent) and TSS (influent) and the outputs were COD and TSS of effluent and the biogas production. In this study beside MLP, Radial Basis Function (RBF) was utilized for prediction. The structure of selected MLP was 4-3-3 and selected RBF was 4-15-3. The average relative error for COD, TSS and biogas for MLP were 8.9%, 15.6% and 6.4% and for RBF were 11.8%, 46.1% and 6.7%, respectively. This study has shown that MLP had the highest predictive power pose (Elnekave et al., 2012).

Shi et al. established a neural network model for studying the effect of input parameters to the performance of Up-flow Anaerobic Sludge Bed Anaerobic Filter Reactor (UASBAF). While the input parameters were COD, HRT, pH and alkalinity, the output was the removal of COD. The structure of the model was 4-3-1 in which tansig and purelin were used as the transfer functions for hidden and output layer, respectively. The training algorithm was traingdx (gradient descent with momentum and adaptive learning rate). The result of this survey about relative contributions was: pH > influent COD > HRT > alkalinity. Moreover, optimum conditions were obtained by applying three-dimensional graphs of affecting factors (Shi et al., 2009).

Parthiban and Parthiban used ANN to model the performance for the treatment of sago wastewater using anaerobic tapered fluidized bed reactor. The input parameters were flow, COD and pH of influent and the outputs were COD of effluent and CH_4 gas. Tansig and purelin were used as the transfer functions for the hidden and output layer, respectively. This study has shown the great adaptability of BP neural network to the variations of system configuration and operation condition. Moreover prediction results (R=0.99) are found to be closer to experimental results (Parthiban and Parthiban 2012).

Sinha et al. studied the feasibility of application of an artificial neural network-based model for simulating the performance of Up-flow anaerobic sludge blanket (UASB) reactors under various conditions of influent wastewater flow and characteristics. The model inputs are organic loading rate (OLR), hydraulic retention time (HRT) and influent bicarbonate alkalinity (BA) and the output variables are one of effluent substrate concentration (S_e), reactor BA, reactor pH, volatile fatty acid (VFA), average gas production rate and percent methane content of the gas. This study has shown promising applications of neural networks for finding the optimum conditions. Moreover, obtained simulation results were found to provide important insights into key variables that were responsible for influencing reactor (Sinha et al., 2002).

Table 2. The results of reviewed models

NO.	Inputs	Output(s)	Structure	Training Algorithm	Reference
1	TS, TVS, pH, T	CH_4	4 – 25 – 25 – 1	BP	(Abu Qdais et al., 2010)
2	Br_{t-1}, Br_{t-2}, pH_{t-1}	pH_t	3 – 3 – 1	BP	(Holubar, et al., 2002)
3	Br_t, Br_{t-1}, Br_{t-2}, pH_{t-1}, GP_{t-1}, GC_{t-1}, VFA_{t-1}, AA_{t-1}, PA_{t-1}	VFA_{tot}, AA_t, PA_t	9 – 3 – 3		
4		GP_t, GC_t	9 – 3 – 2		
5	$(SO_4^{2-})_{t-1}$, $(SO_4^{2-})_t$, $(SO_4^{2-})_{t+1})$, COD_t, COD_{t+1}, H_2S_t, H_2S_{t-1}	H_2S_{t+1}	7 – 5 – 1	Batch gradient descent with momentum algorithm	(Strik et al., 2005)
6	N_{t+1}, COD_{t+1}, (biogas productivity)$_t$, pH_t, pH_{t-1}, pH_{t-2}, NH_{3t}, N-NH_{3t}	N-NH_{3t+1}	8 – 7 – 1		
7	pH, alkalinity, conductivity, cl^-, COD, SO_4^{2-} of leachate, waste temperature, refuse age	CH_4 fraction with leachate recirculation	8 – 12 – 1	L – M	(Ozkaya et al., 2007)
8		CH_4 fraction without leachate recirculation	8 – 15 – 1		
9	SRT+SI+VDS+FCOD	BMP	-	L – M	(Kianmehr et al., 2014)
10	SRT+SI+VDS+FTKN				
11	SRT+SI+FCOD+FTKN				
12	SRT+SI+VDS				
13	SRT+SI+FCOD				
14	SI+VDS+FCOD				
15	SRT+SI+VDS+FCOD+ FTKN				
16	Q, T, pH, influent-VFA, alkalinity	CH_4	6 – 13 – 1	L – M	(Güclü et al., 2011)
17			5 – 2 – 1		
18			6 – 11 – 3 – 1		
19			5 – 13 – 2 – 1		
20	Q, T, pH, influent-VFA alkalinity, influent-VS	VS	6 – 35 – 1	GD	
21			5 – 4 – 1		
22			6 – 11 – 2 – 1		
23			5 – 8 – 2 – 1		

Table 2. (Continued)

NO.	Inputs	Output(s)	Structure	Training Algorithm	Reference
24	OLR, VFA, alkalinity of influent and effluent, pH of influent and effluent, T	Biogas Production			(Kanat and Saral 2009)
25	CA, CB_1, CB_2	CH_4	3 – 8 – 1	L – M	(Dong and Zhao 2013)
26		CO_2	3 – 11 – 1		
27		total gas	3 – 2 – 1		
28	OLR, influent-pH	COD, biogas, VFA, alkalinity, effluent-pH	-	BP	(Parthiban et al., 2007)
29	CD, BS, RB, PW, SD	Biogas performance index of digester	5 – 2 – 1	L – M	(Kana et al., 2012)
30	COD(influent), TSS, OLR, T, VFA(influent)	COD and VFA (effluent)	6 – 12 – 2 – 2	GD	(Badalians Gholikandi et al., 2014a)
31	Flow, Volumetric load, COD_{in} and TSS_{in}	COD_{out}, TSS_{out} and Biogas Production	4 – 3 – 3	L – M	(Elnekave et al., 2012)
32			4 – 15 – 3		
33	Last eight BA data points	Likelihood of being out of control	8 – 8 – 1	BP	(Wilcox et al., 1995)
34	COD, HRT, pH and alkalinity	Removal of COD	4 – 3 – 1	Gradient descent with momentum and adaptive learning rate	(Shi et al., 2009)
35	Flow, COD and pH of influent	COD of effluent and CH_4 gas	-	BP	(Parthiban and Parthiban 2012)
36	COD of influent (under six different volumetric organic loads)	COD of effluent	-	Batch gradient descent with momentum algorithm	(Gasim et al., 2013)
37	OLR, HRT, 132influent-BA,	S, BA, pH, VFA, gas prediction rate, methane content	-	-	(Sinha et al., 2002)
38	T, pH, TS, VS, sludge flow rate, OL, time	Biogas production	-	-	(Wei and Kusiak 2012)

Gasim et al. developed neural network model for simulating Up-flow anaerobic sludge blanket (UASB) reactor which was used for treating petroleum refinery wastewater under six different volumetric organic loads. In this research two different transfer functions were tested for the hidden layer while the output layer one was chosen purelin. The selected network which was trained by batch gradient descent with momentum (traingdm) had 12 hidden neurons and utilized tansig and purelin as the transfer functions. The results of this simulation were: $R^2 = 0.866$ and RMSE = 0.074. By using the optimum model with random data entries ranged between 500 and 9000 resulted in a pattern that simulates the reactor performance for data that were never really experimentally tested in the lab (Gasim et al., 2013).

Wei and Kusiak applied MLP neural network to construct the model for optimization of biogas production in a wastewater treatment. This study has shown that MLP offers better prediction accuracy than the other models. The input parameters were temperature (T), total solids (TS), volatile solids (VS), pH, sludge flow rate, organic load (OL) and detention time and the output was biogas production. The mean absolute percentage error of the constructed model was 0.07 and the index of agreement of 0.99 indicated high correlation between the predicted and observed values. Moreover, for optimizing biogas production, a single-objective optimization model was formulated and solved with a particle swarm optimization (PSO) (Wei and Kusiak 2012).

4. Implementing MLP and SOFM Neural Networks

4.1. MLP Neural Network

For proposing more details of these kinds of neural networks, we have modeled data which were gathered from a research by Valladao et al. (Valladao et al., 2011). These data are shown in table 4. For implementing the MLP neural network, Matlab™ software was utilized. For this target different neural networks were created, trained by L-M (trainlm) and tested. Numbers of hidden neurons were varied from 1 to 25. Moreover, tansig and logsig were used for hidden layer and beside these two activation functions; linear (purelin) was also used for output layer. In order to overcome over-fitting issue, early stopping method was applied in which 60% of data were used for training, 20% for validation and 20% for testing. Since variables have different magnitude, they cannot be fed directly to the MLP. Otherwise process variables with larger magnitude yield unequal importance due to the nature of the weight update procedure (Cinar 2005). Therefore, all data fed to the network have been preprocessed in which have been converted into ranges of [-1, 1]. The best model structure which has the lowest MSE (MSE = 38.77) was obtained of 3:20:1 structure and is shown in Figure 3. The activation function of this model is tansig for hidden layer and purelin for output layer which is a common choice for function approximation neural networks. The R value of 0.938 for training data, 0.99 for validation and 0.91 for test subsets show high performance of the chosen model. Figure 4 shows correlations between predicted and measured (experimental) methane production for data subsets. All these processes have been done by using Matlab™ scripts.

Table 4. The experimental data used for simulation

Hydrolysis time (hr)	COD (in) mg/l	COD (out) mg/l	Time (day)	Volume methane (ml)
4	8775	238	12	100.4
8	9412	173	12	102
24	7087	143	12	71.4
4	7166	142	7	72.2
8	7250	128	7	46.6
24	7525	142	7	73.4
4	3350	149	9	113.3
8	3937	128	9	77
24	5160	145	9	97.8
4	8775	238	12	104.7
8	9412	173	12	119.8
24	7087	143	12	75.7
4	7166	142	7	76.1
8	7250	128	7	52.6
24	7525	142	7	82.3
4	3350	149	9	142.2
8	3937	128	9	85.7
24	5160	145	9	98.7

Figure 5 shows a plot of training, validation and test errors as shown that the MSE of these subsets are decreasing through training the network and figure 6 shows the performance of network and it indicates that selected network outputs are approximately near to the targets.

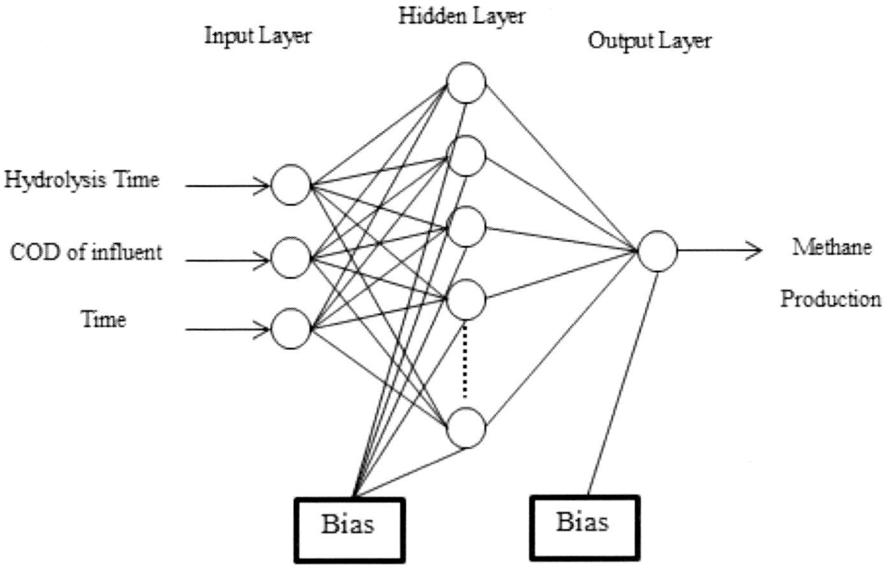

Figure 3. The structure of MLP.

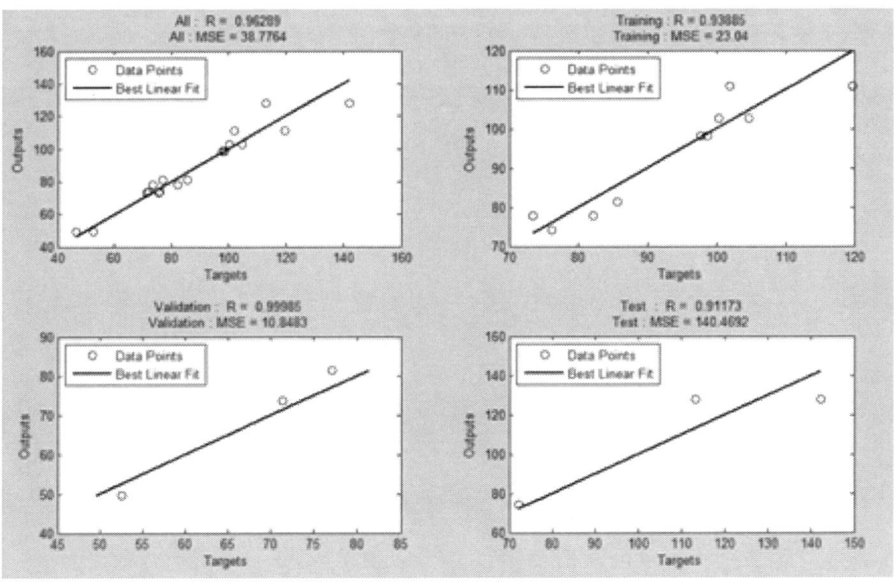

Figure 4. Correlations between predicted and measured (experimental) methane production for the data subsets.

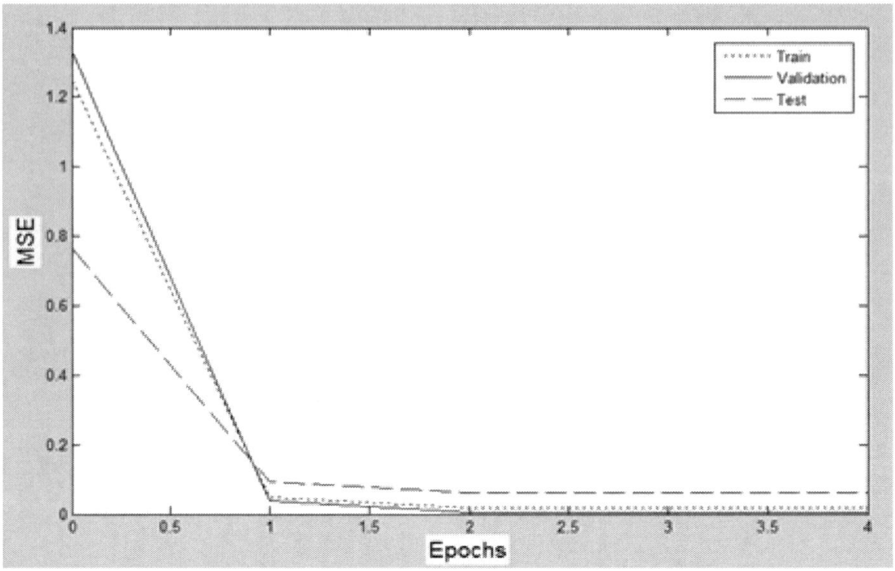

Figure 5. Decreasing MSE of the data subsets.

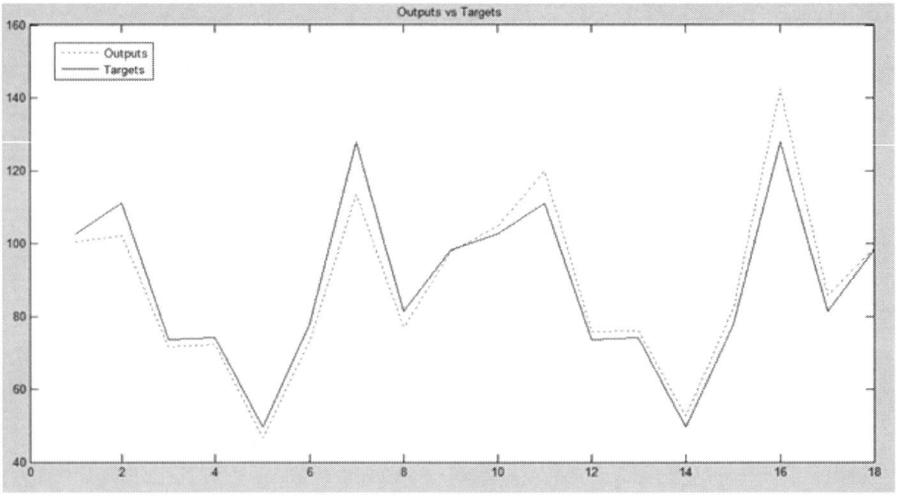

Figure 6. Targets vs. Outputs.

For finding the relative importance of input variables on methane production, Garson equation (Eq. 7) was applied which uses the weights of the best trained network (Elmolla et al., 2010).

$$I_j = \frac{\sum_{m=1}^{m=Nh}((\left|W_{jm}^{ih}\right| \sum_{k=1}^{Ni}\left|W_{km}^{ih}\right|) \times \left|W_{mn}^{ho}\right|)}{\sum_{K=1}^{k=Ni}[\sum_{m=1}^{m=Nh}((\left|W_{km}^{ih}\right| / \sum_{k=1}^{Ni}\left|W_{km}^{ih}\right|) \times \left|W_{mn}^{ho}\right|)]} \quad (7)$$

Where I_j is the relative importance of the j_{th} input variable on the output variable, N_i and N_h are the number of input and hidden neurons and the W is for connection weights, respectively. The superscripts $'i'$, $'h'$ and $'o'$ are for input, hidden and output layers, respectively and subscripts $'k'$, $'m'$ and $'n'$ refer to input, hidden and output neurons, respectively (Delnavaz et al., 2010). Table 5 shows the best ANN model weights which were used in equation 7.

Table 5. Weights between Input & Hidden Layers (W1) and between Hidden & Output Layers (W2)

	W1				W2	
	Input Variables				Output	
Neuron	Hydrolysis time (h)	COD (in) mg/l	Time (day)	Bias	Neuron	Volume methane (ml)
1	2.0292	2.3437	-2.1979	0.7672	1	-3.8002
2	3.5405	1.2155	-0.6546	-0.151	2	-3.4002
3	3.5002	1.2038	0.8609	0.2145	3	-3.0001
4	0.5126	1.273	3.5437	-0.859	4	-2.6001
5	0.6336	2.4644	-2.8225	0.8495	5	-2.2001
6	-2.493	2.6383	-1.1251	0.2842	6	1.8001
7	0.2214	0.2697	-3.7841	-0.791	7	-1.4001
8	0.4507	-3.2372	-1.9388	0.4005	8	-1.0000
9	-2.6913	-0.9595	-2.5055	-0.208	9	0.6000
10	-0.0404	2.1212	-3.1528	-0.83	10	0.2000
11	0.3991	-3.7352	-0.5751	-0.571	11	0.2000
12	-2.6586	-1.5008	-2.2629	-0.502	12	-0.6000
13	0.7708	2.9193	2.3076	-0.547	13	1.0000
14	-3.1871	2.0608	0.1925	0.406	14	-1.4001
15	-1.2979	-1.9897	2.9661	0.5083	15	-1.8001
16	0.1819	2.7543	2.6119	0.0946	16	2.2001
17	-1.3798	-3.4749	0.6804	0.107	17	-2.6001
18	0.9673	-0.7054	-3.6067	0.2611	18	3.0001
19	-1.1446	-0.793	3.5359	0.9709	19	-3.4002
20	-3.5588	-0.9719	0.912	0.2686	20	-3.8002
					Bias	0.2009

The relative importance of different variables of the process is shown in figure 7 which shows the higher effect of time on methane production.

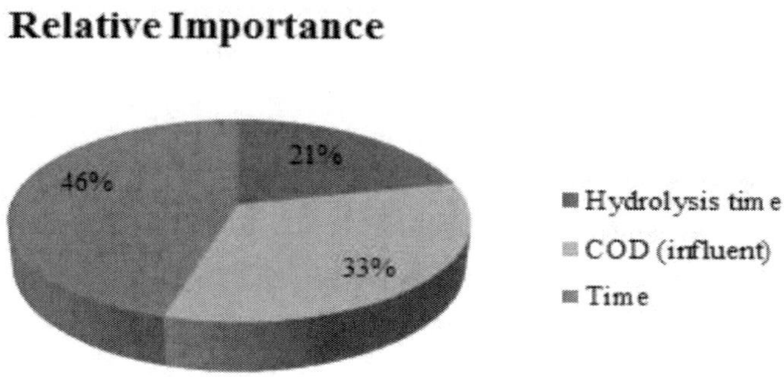

Figure 7. The relative importance of the variables.

For more investigation of time influence on methane production, the selected MLP neural network was evaluated, whereas two parameters (COD (influent) and Hydrolysis Time) were changed simultaneously and another one (Time) was considered constant (Badalians Gholikandi et al., 2014a). Therefore, two figures (figure 8 and 9) were obtained in two different times (9 and 12 (day))

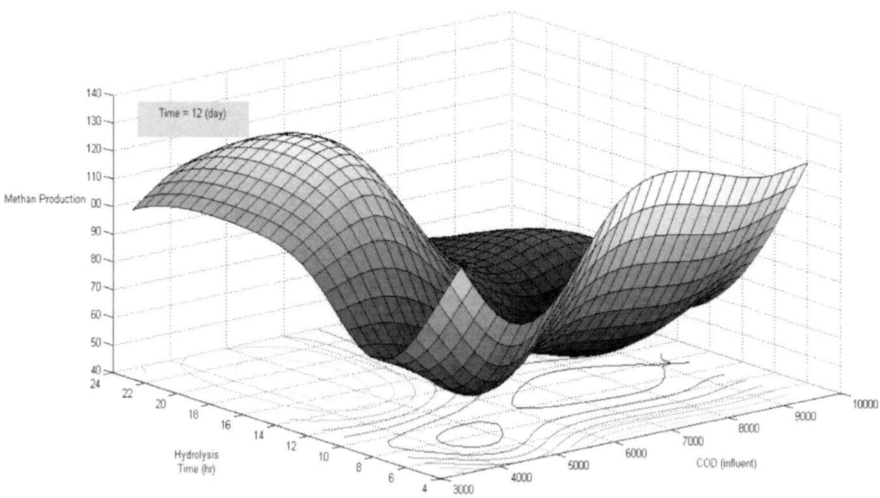

Figure 8. The simultaneous effect of Hydrolysis Time & COD on Methane Production (Time=12 days).

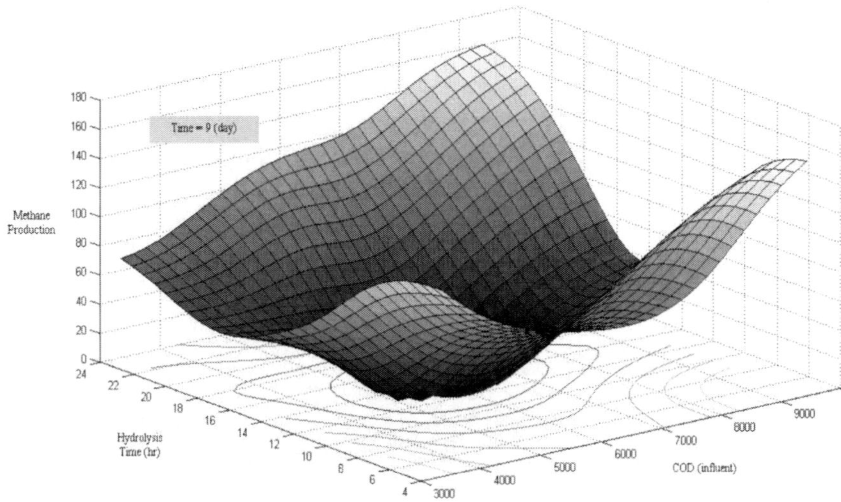

Figure 9. The simultaneous effect of Hydrolysis Time & COD on Methane Production (Time=9 days).

In the figure 9, when the hydrolysis time is 4 – 6 hours, methane production will increase by enhancing COD of influent. But in the range of 10 – 15 hours of hydrolysis time, methane production will decrease. Also, by increasing hydrolysis time from 16 to 25 hours and enhancing COD (influent), methane volume will increase and it reaches to its maximum level (150 ml). By applying this figure, methane production can be estimated in specified span even not examined range.

4.2. SOFM Neural Network

For finding regularities and correlations in the input variables, SOM Toolbox (www.cis.hut.fi/projects/somtoolbox/) is selected and has been used with MATLAB^{TM}. Due to different magnitudes of inputs, they cannot be used without normalization. For preprocessing, all variables were scaled which have the variance of 1 and the mean of 0 (Badalians Gholikandi et al., 2014a). The map size during training was 5×4 that is sufficient regarding to Eq.6 and final quantization and topographic error were 0.983 and 0.056, respectively. The correlation between the studied variables has been shown by component plane visualization in figure 10. The most important features of SOFM are its visualization capabilities (Hong et al., 2003) in which each component plane (each hexagon) represents one map and its colors tells the value of the component in that node (Hong et al., 2003; Grieu et al., 2006; Rustum et al., 2008). For each variable, there is a component planes (Vesanto 1999; Rustum et al., 2008) in which hexagons correspond to the same map node (Hong et al., 2003). Each component plane has a legend on its right side which indicates the range of each variable.

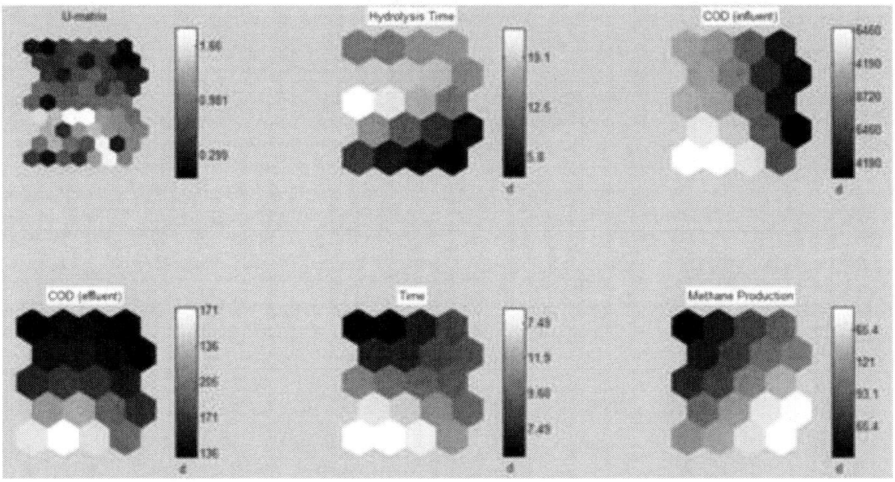

Figure 10. The results of visualization of the SOFM.

By considering figure 10 it can be indicated that maximum methane production will be obtained in conditions which are expressed in table 6. By comparing the tables 4 and 6, the results acquired by SOFM neural network (table 6) have good corresponding to experimental data (Tab. 4). Moreover, the figure 10 presents shows that the most effective parameter is the retention time which was also obtained in relative importance (Fig. 7).

Table 6. The SOFM results for maximum methane production

Parameters	SOFM Results
Hydrolysis time (h)	5.8-9.1
COD (in) mg/L	5000-8720
COD (out) mg/L	160-205
Time (day)	5.8-9.5
Volume methane (mL)	100-121

Conclusion

The challenge of global pollution and depletion of fossil fuels has made modern society to find alternative renewable sources (Kana et al., 2012). Methane as a valuable energy source is the main component of the biogas which is produced by anaerobic digestion process that has been widely used for the treatment of primary and secondary sludge in wastewater treatment plants, animal manure, organic wastes from households and industries (Levstek and Lakota 2010). The main advantages of this process include less biomass sludge production, more effective pathogen removal, minimal odor emissions, applicable for high-strength industrial wastes and lower operating costs due to utilizing CO_2 in spite of O_2 (Apples et al., 2008; Ward et al., 2008). The optimization of anaerobic digestion process due to its complexity and considering different operating conditions can be pursued by using appropriate digestion models (Apples et al., 2008). Moreover, the modeling will also be used to develop, apply and validate digestion on-line monitoring (Apples et al., 2008).

Establishing a model can be done in two ways: (1) Mechanistic models that include deriving in a deductive manner by using the laws of nature and (2) Black-Box modeling which is evolved from several experiments (Strik et al., 2005). Because of complexity of anaerobic digestion process, there isn't a well-understanding of parameters or parameters determination. Therefore, the black-box modeling is selected (Strik et al., 2005). The most popular widely used black-box modeling approach is artificial neural network (ANN). In recent years this kind of modeling approach is gaining more attention because of its adaptability, generalization and robustness. The ANN has been used in different environmental areas such as hydrology, water quality, water treatment, municipal and industrial wastewater treatment plants and etc. One of the aforementioned areas is anaerobic treatment process. Many researchers have utilized the ANN in anaerobic digestion processes, especially methane production. These investigations have been proposed encouraging estimation results for online control in both steady state and abnormal conditions. Moreover, these models have identified the nonlinear relationship between the various experimental parameters. In these studies the MLP method had the highest predictive power pose. Elnekave et al. have shown that the MLP method is more precise than the RBF (Elnekave et al., 2012). This review attempts to summarize the efforts that have been accomplished in simulating anaerobic digestion process and highlights parameters selection, training algorithm and structure of applied networks. In this work we also applied

SOFM beside MLP neural network. The SOFM method was utilized to find regularities and correlations in input variables and also the range that indicates maximum methane production. Moreover, the MLP method was employed to investigate the time influence on methane production in which two parameters, i.e., COD of influent and Hydrolysis Time were changed simultaneously and Time was considered constant. For finding the relative importance of input variables on methane production, Garson equation (Eq. 7) was applied which uses the weights of the best trained network.

It can be concluded that the artificial neural networks modeling has the capability to be used as a decision support system (DDS) to optimize system operation (Holubar et al., 2002).

REFERENCES

Abu Qdais, H., Bani Hani K. & Shatnawi N. (2010). Modeling and optimization of biogas production from a waste digester using artificial neural network and genetic algorithm. Resources. *Conservation and Recycling, 54(6)*, 359-363.

Al Seadi T. (2009). Good practice in Quality Management of AD Residues. *University of Southern Denmark Esbjerg,* 3-4.

Appels L., Baeyens J., Degrève J. & Dewil R. (2008). Principles and potential of the anaerobic digestion of waste-activated sludge. *Progress in Energy and Combustion Science 34(6),* 755-781.

Badalians Gholikandi G. (2009). Wastewater Treatment Plant Design, Ayeej Press: Tehran, Iran.

Badalians Gholikandi G., Khosravi M. (2012a). Upgrading of submerged membrane bioreactor operation with regard to soluble microbial products and mathematical modeling for optimization of critical flux. *Journal Desalination and Water Treatment ,39,*199–208.

Badalians Gholikand G., Hazrati H. & Rostamian H. (2012b). An Artificial Neural Network Model for Predict Performance of Pressure Filters and Determining Optimum Turbidity for Removing Coli-form and Total Bacteria. *Journal of Environmental Studies, 37,* 129-136.

Badalians Gholikandi G., Alibeygi MR. & Rahmani S. (2011a). Prediction of biological UCT process efficiency using artificial neural network. *International Journal Wastewater Treatment Green Chem., 2,* 45–55.

Badalians Gholikandi G., Delnavaz M. & Riahi R. (2011b). Use of Artificial Neural Network for Prediction of Coagulation/Flocculation Process by

PAC in Water Treatment Plant. *Environmental Engineering and Management Journal ,10(11),* 1719-1725.

Badalians Gholikandi G., Masihi H., Azimipour M., Abrishami A. & Mirabi M. (2014a). Optimizing stabilization of waste-activated sludge using Fered-Fenton process and artificial neural network modeling (KSOFM, MLP). *Journal Environmental Science and Pollution Research.*

Badalians Gholikandi G., Jamshidi S. & Hazrati H. (2014b). Optimization of Anaerobic Baffled Reactor (ABR) Using Artificial Neural Network in Municipal Wastewater Treatment. *Journal of Environmental Engineering and Management ,13(1),* 95-104.

Boe K. (2006). Online monitoring and control of the biogas process. Ph.D. Thesis, Institute of Environment & Resources, Technical University of Denmark.

Bitton G. (2005). Wastewater microbiology. John Wiley & Sons.

Blumensaat F, Keller J. (2005). Modelling of two-stage anaerobic digestion using the IWA Anaerobic Digestion Model No. 1 (ADM1). *Water Research ,39,* 171-183.

Boone D.R., Xun L.Y. (1987). Effects of pH, temperature, and nutrients on propionate degradation by a methanogenic enrichment culture. *Applied and Environmental Microbiology ,53,* 1589–1592.

Cinar O. (2005). New tool for evaluation of performance of wastewater treatment plant: Artificial neural network. *Journal of process biochemistry, 40,* 2980–2984.

DaelsT., Willems B., Vervaeren H., Dejans P., Maes G., Dumoulin A. & Van Hulle S.W.H. (2009). Calibration and statistical analysis of a simplified model for the anaerobic digestion of solid waste. *Environmental Technology ,30,* 1575-1584.

Delnavaz M., Ayati B. & Ganjidoust H. (2010). Prediction of moving bed biofilm reactor (MBBR) performance for the treatment of aniline using artificial neural networks (ANN). *Journal of Hazardous Materials ,179,* 769–775.

Derbal K., Bencheikh-Iehocine M., Cecchi F., Meniai A. H. & Pavan P. (2009). Application of the IWA ADM1 model to simulate anaerobic co-digestion of organic waste with waste activated sludge in mesophilic condition. *Bioresource Technology ,100,* 1539-1543.

Djebbar Y., Narbaitz R.M. (2002). Neural network prediction of air stripping KLa. *Journal of Environmental Engineering ,128 (5),* 451–460.

Dochain D., Vanrolleghem P. (2001). Dynamical Modelling and Estimation in Wastewater Treatment Processes Control of Bioreactors. *IWA Publ.,* UK.

Dochain D. (1995). Recent approaches for the modeling, monitoring and control of anaerobic digestion processes. *In Proceedings of the international workshop on monitoring and control of anaerobic digestion processes*, 6-7.

Donoso-Bravo A., Mailier J., Martin C., Rodríguez J., Aceves-Lara C.A. & Wouwer A.V. (2011). Model selection, identification and validation in anaerobic digestion: a review. *Water Research ,45(17),* 5347-5364.

El-Hinnawi E., Biswas A.K. (1981). Renewable sources of energy and the environment. *Natural resources and the environmental series 6, 1st ed.*

Elmolla E.S., Chaudhuri M. & Eltoukhy M.M. (2010). The use of artificial neural network (ANN) for modeling of COD removal from antibiotic aqueous solution by the Fenton process, *Journal of Hazardous Materials, 179,* 127–134.

Elnekave M., Celik S.O., Tatlier M. & Tufekci N. (2012). Artificial Neural Network Predictions of Up-Flow Anaerobic Sludge Blanket (UASB) Reactor Performance in the Treatment of Citrus Juice Wastewater. *Polish Journal of Environmental Studies, 21(1).*

Garcia H., Gonzalez L. (2004). Self-Organizing Map and Clustering for Wastewater Treatment Monitoring. *Eng. Appl. Artificial Intelligence ,17 (3),* 215-225.

Gasim H.A., Kutty S.R.M., Hasnain Isa M. & Alemu L. T. (2013). Optimization of Anaerobic Treatment of Petroleum Refinery Wastewater Using Artificial NeuralNetworks. *Research Journal of Applied Sciences, Engineering and Technology ,6(11),* 2077-2082.

Gerardi M.H. (2003). The microbiology of anaerobic digesters. John Wiley & Sons.

Gerardi M.H. (2006). Wastewater bacteria (Vol. 5). John Wiley & Sons.

Güclü D., Yilmaz N. & Ozkan-Yucel U.G. (2011). Application of neural networks prediction model to full-scale anaerobic sludge digestion. *Journal of Chem. Technol. Biotechnology, 86,* 691-698.

Grieu S., Thiery F., Traore A., Nguyen T.P., Barreau M. & Polit M. (2006). KSOM and MLP neural networks for on-line estimating the efficiency of an activated sludge process. *Journal of Chem. Engineering, 116,* 1–11.

GTZ (editor). (2009). Biogas sanitation for blackwater or brownwater, or excreta treatment and reuse in developing countries. Eschborn: German agency for technical cooperation (GTZ) GmbH and sustainable sanitation alliance (SuSanA).

Hamed M.M., Khalafallah M.G. & Hassanien E.A. (2004). Prediction of wastewater treatment plant performance using artificial neural networks. *Environmental Modelling & Software ,19(10),* 919-928.

Haykin S. (1994). Neural networks: a comprehensive foundation. Prentice Hall PTR.

Hernandez S.C., Bueno J.A., Sanchez E.N. & Diaz-Jimenez L. (2013). State Estimation by Artificial neural networks in a continuous bioreactor. *12th IFAC Symposium on Computer Applications in Biotechnology*, 16-18.

Hill D.T., Barth C.L. (1977). A Dynamical Model for Simulation of Animal Digestion. *Journal of Water Pollution Contr. Federation, 10,* 2129-2143.

Holubar P., Zani L., Hager M., Froeschl W., Radak Z. & Braun R. (2002). Advanced controlling of anaerobic digestion by means of hierarchical neural networks. *Water Research, 36*, 2582-2588.

Hong Y.S., Rosen M.R. & Bhamidimarri R. (2003). Analysis of a municipal wastewater treatment plant using a neural network-based pattern analysis. *Water Research, 37,* 1608-1618.

Hwang M.H., Jang N.J., Hyum S.H. & Kim I.S. (2004). Anaerobic bio-hydrogen production from ethanol fementation: the role of pH. *Journal of Biotechnology, 111,* 297–309.

Kana E.B.G., Oloke J.K., Lateef A. & Adesiyan M.O. (2012). Modeling and optimization of biogas production on saw dust and other co-substrates using Artificial Neural network and Genetic Algorithm. *Journal of Renewable Energy ,46,* 276-281.

Kanat G., Saral A. (2009). Etimation of biogas production rate in a thermophilic UASB reactor using artificial neural networks, *Journal of Environ. Model. Assessment ,14,* 607-614.

Kleerebezem R., Van Loosdrecht M.C.M. (2006). Waste characterization for implementabon in ADM1. *Water Science and Technology ,54,* 167-174.

Khataee A.R., Kasiri M.B. (2010). Artificial neural networks modeling of contaminated water treatment processes by homogeneous and heterogeneous nanocatalysis. *Journal of Molecular Catalysis A: Chemical, 331(1),* 86-100.

Kianmehr P., Masnsoor W. & Kfoury F.A. (2014). Prediction of biogas generation profiles in wastewater treatment plants using neural network. *Journal of Clean Energy Technologies, 2(3),* 201-205.

Lettinga G. (1995). Anaerobic digestion and wastewater treatment systems. *Antonie van Leeuwenhoek ,67,* 3– 8.

Levstek T., Lakota M. (2010). The use of artificial neural networks for compounds prediction in biogas from anaerobic digestion-A review. *Journal of AGRICULTURA, 7,* 15-22.

Lubken M., Wichern M., Schlattmann M., Gronauer A. & Horn H. (2007). Modelling the energy balance of an anaerobic digester fed with cattle manure and renewable energy crops. *Water Research, 41,* 4085-4096.

Maghanaki M., Mohammadi M., Ghobadian B., Najafi G. & Janzadeh Galogah R. (2013). Potential of biogas production in Iran. *Renewable and Sustainable Energy Reviews, 28,* 702-714.

Mackie R.I., Bryant M.P. (1981). Metabolic activity of fatty acid-oxidizing bacteria and the contribution of acetate, propionate, butyrate and CO_2 to methanogenesis in cattle waste at 40 and 60°C. *Applied Environ. Microbiology ,41,* 1363–1373.

Mata-Alvarez J. (2002). Biomethanization of the Organic Fraction of Municipal Solid Wastes, IWA Publishing.

Mara D., Horan N.J. (Eds.). (2003). Handbook of water and wastewater microbiology. Academic press.

McInernay M.J., Bryant M.P., Hespell R.B. & Costerton J.W. (1981). Syntrophomonas wolfei, gen. nov. sp. nov., an anaerobic syntrophic, fatty acid-oxidizing bacterium. *Applied Environ. Microbiology ,41,* 1029–1039.

Moletta R., Verrier D. & Albagnac G. (1986). Dynamic modeling of anaerobic digestion. *Water Research, 20(4),* 427- 437.

Molina F., Castellano M., Garcia C., Roca E. & Lema J.M. (2009). Selection of variables for on-line monitoring, diagnosis, and control of anaerobic digestion processes. *Water Science & Technology ,60(3).*

Ozkaya B., Demir A. & Bilgili M.S. (2007). Neural network prediction model for the methane fraction in biogas from field-scale landfill bioreactors. *Environmental Modeling & Software, 22,* 815-822.

Parkin G.F., Owen W.F. (1986). Fundamentals of anaerobic digestion of wastewater sludges. *Journal of Environmental Engineering, 112 (5),* 867-1120.

Parthiban R., Iyer P. & Sekaran G. (2007). Anaerobic tapered fluidized bed reactor for starch wastewater treatment and modeling using multilayer perceptron neural network. *Journal of Environmental Sciences ,19,*1416-1423.

Pendashteh A.R., Fakhru'l Razi A., Chaibakhsh N., Chuah Abdullah L., Madaeni S.S. & Zainal Abidin Z. (2011). Modeling of membrane bioreactor treating hypersaline oily wastewater by artificial neural network. *Journal of Hazardous Materials, 192,* 568– 575.

Pipoli T. (2005). Feasibility of biomass-based fuel cells for manned space exploration, In: *Proceedings of the seventh European space power conference,* Stresa, Italy.

Polprasert C. (1989). Organic Wastes Recycling. John Wiley & Sons, Chichester, U.K.

Pullammanappallil P.C., Svoronos S.A., Chynoweth D.P. & Lyberatos G. (1998). Expert system for control of anaerobic digesters. *Biotechnology and Bioengineering 58,* 13–22.

Rahmanian B., Pakizeh M., Mansoori S.A.A. & Abedini R. (2011). Application of experimental design approach and artificial neural network (ANN) for the determination of potential micellar-enhanced ultrafiltration process. *Journal of Hazardous Materials, 187,* 67–74.

Rumelhart D.E., Hinton G.E. & Williams R.J. (1986). Learning representations by back-propagating errors. *Nature ,323,* 533 – 536.

Rustum R., Adeloye A.J. & Scholz M. (2008). Applying Kohonen Self-Organizing Map as a Software Sensor to Predict Biochemical Oxygen Demand. *Water Environment Research, 80,* 32-40.

Salari D., Daneshvar N., Aghazadeh F. & Khataee A.R. (2005). Application of artificial neural networks for modeling of the treatment of wastewater contaminated with methyl tert-butyl ether (MTBE) by UV/H_2O_2 process. *Journal of Hazardous Materials, 125,* 205–210.

Sahm H. (1984). Anaerobic wastewater treatment. Springer Berlin Heidelberg, 84-115.

Sarle W.S. (2005). Neural network faq, 2002. Official FAQ of the Usenet newsgroup comp. ai. neural-nets (ftp://ftp. sas. com/pub/neural/FAQ. html).

Savegnago R.P., Nunes B.N., Caetano S.L., Ferraudo A.S., Schmidt G.S., Ledur M.C. & Munari D.P. (2011). Comparison of logistic and neural network models to fit to theegg production curve of White Leghorn hens", *Pollution Science, 90,* 705–711.

Shi Y., Zhao X.T., Zhang Y.M. & Ren N.Q. (2009). Back propagation neural network (BPNN) prediction model and control strategies of methanogen phase reactor treating traditional Chinese medicine wastewater (TCMW). *Journal of Biotechnology ,144(1),* 70-74.

Sinha S., Bose P., Jawed M., John S. & Tare V. (2002). Application of neural network for simulation of upflow anaerobic sludge blanket (UASB) reactor performance. *Biotechnology and bioengineering, 77(7),* 806-814.

Speece R.E. (1983). Anaerobic biotechnology for industrial wastewater treatment. *Environmental Sci. Technology, 17,* 416A–427A.

Steyer JP; Amouroux M; Moletta R. (1995). Process modeling and control to improve stable operation and optimization of anaerobic digestion processes. In: *Proceedings of the International Workshop on Monitoring and Control of Anaerobic Digestion Processes, 6–7/12/1995,* Laboratoire de Biotechnologie de L'environnement INRA, France, 30–5.

Strik D.P.B.T.B., Domnanovich A.M., Zani L., Braun R. & Holubar P. (2005). Prediction of trace compounds in biogas from anaerobic digestion using the MATLAB neural network toolbox. *Environmental Modeling & Software, 20,* 803-810.

Tobiszewski M., Tsakovski S., Simeonov V. & Namieśnik J. (2012). Chlorinated solvents in a petrochemical wastewater treatment plant: An assessment of their removal using self-organising maps. *Chemosphere ,87(8),* 962-968.

Turovskiy I.S., Mathai P.K. (2006). Wastewater sludge processing. John Wiley & Sons.

Valladao A.B.G., Torres A.G., Freire D.M.G. & Cammarota M.C. (2011). Profiles of fatty acids and triacylglycerols and their influence on the anaerobic biodegradability of effluents from poultry slaughterhouse. *Journal of Bioresource Technology, 102,* 7043-7050.

Vesanto J. (1999). SOM-based data visualization methods. *Intelligent data analysis, 3(2),* 111-126.

Ward A.J., Hobbs P.J., Holliman P.J. & Jones D.L. (2008). Optimisation of the anaerobic digestion of agricultural resources. *Journal of Bioresource Technology ,99(17),* 7928-7940.

Wei X., Kusiak A. (2012). Optimization of Biogas Production Process in a Wastewater Treatment Plant. *In Proceedings of Industrial and Systems Engineering Research Conference.*

Wilcox S.J., Hawkes D.L., Hawkes F.R. & Guwy A.J. (1995). A neural network based on bicarbonate monitoring to control anaerobic digestion. *Water Research, 29 (6),* 1465-1470.

Yordanova S., Petrova R. & Mladenov V. (2004). Neuro-fuzzy control for anaerobic wastewater treatment. *WSEAS Transactions on Systems, 2(3),* 724-729.

Yu L., Wensel P.C., Ma J. & Chen S. (2013). Mathematical modeling in Anaerobic Digestion (AD). *Journal of Bioremed. Biodegradations 4.*

In: Methanogenesis
Editor: Gagik Badalians Gholikandi
ISBN: 978-1-63321-567-2

Chapter 6

THE ANAEROBIC SLUDGE DIGESTION PROCESS

Hamidreza Masihi, Gagik Badalians Gholikandi* and Mohammad Azimipour
Faculty of Water Engineering and Environment,
Shahid Beheshti University, A. C., Tehran, Iran

ABSTRACT

The common preferred process for the waste sludge stabilization is the anaerobic digestion, due to related energy recovery from sludge, i.e., biogas production and anaerobically-digested sludge application, i.e., agricultural and industrial uses. Consequently, the environmental problem caused by sewage sludge can be solved in a sustainable way. The bacterial process occurs when organic matter decomposes biologically and transforms into biogas in an environment devoid of oxygen. In this chapter, emphasis is mainly put on stages of process occurrence, significant affecting factors, inhibitory substances and process improving methods, i.e., pretreatment methods.

Keywords: anaerobic digestion, sludge stabilization, abiotic factors, inhibitory substances, pretreatment

---

* Corresponding author: Hakimieh- Tehranpars, Shahid Abbaspour Blvd., PO Code 16589-53571, PO Box 16765-1719, Tehran, Iran. Email: g.badalians@yahoo.com.

1. INTRODUCTION

Anaerobic digestion process is the most common process for stabilizing wastewater sludge containing organic matter. Anaerobic decomposition of the organic matter creates considerably less biomass than the aerobic process. The degradation of organic molecules occurs without free molecular oxygen. An anaerobic sludge digester is designed to encourage the growth of anaerobic bacteria and methane-forming archaea bacteria that decreases existing organic matter by reducing them to soluble substances and biogas, mostly carbon dioxide and methane. Abiotic factors, i.e., the nonliving components and operational conditions also affect the activity and growth of the biomass.

The process has been described by four steps, i.e., hydrolysis, fermentation, acetogenesis, and methanogenesis. Four different groups of microorganisms have a symbiotic relationship. The stability of the sludge digestion process relies on the proper balance of the four biological process steps. There are two main pathways of methanogenesis process, i.e., the splitting of acetate by aceticlastic, methane-forming bacteria and the reduction of carbon dioxide by hydrogen-oxidizing, methane-forming bacteria that occur in an anaerobic digester. The hydrogen gas utilization by methane gas producing process reduces the hydrogen gas to a low pressure in the anaerobic digester. Consequently, the acetogenic bacteria are able to produce acetate. Acetoclastic methane-producing bacteria metabolize acetate and produce methane and carbon dioxide. The hydrogenotrophic methane-producing bacteria also combine the carbon dioxide with hydrogen to produce methane (Gerardi, 2006; Badalians Gholikandi, 2009).

Hydrolysis is the process rate limiting step of the sludge digestion process at a lower temperature than 20^0C (van Haandel et al., 1994; Borja et al., 1995). Anaerobic hydrolytic bacteria solubilize complex organic compounds to soluble substrates. A wide range of pH equal 5.0 to 11.0 can be determined as an allowable condition for the hydrolytic bacteria (Stronach et al., 1986). Product inhibition is one of the observed phenomena during hydrolysis occurrence. Substrate and product concentrations in the digester liquor can partly influence the synthesis of extracellular enzymes by the hydrolytic bacteria (Stronach et al., 1986). According to different observations in this regard, high concentration of hydrolysis products inhibits the formation of extracellular enzymes (Fogarty and Kelly, 1979). Consequently, an accumulation of hydrolysis end products can slow down the hydrolysis occurrence, which may eventually lead to a breakdown of the digestion process.

Acidogenesis consists of breaking down the products of hydrolysis performed by fermentative bacteria and results in the production of a variety of simple, soluble organic compounds, most prominently volatile fatty acids, such as butyric and propionic acids, along with H_2, CO_2, and acetic acid (Santosh et al., 2004). Therefore, fermentative bacteria also are known as acid-forming bacteria. Facultative and obligate anaerobic bacteria perform the acidogenesis process. The presence of a large number of facultative anaerobes is important because they can metabolize through the oxidative pathway. Consequently, they remove any existing dissolved oxygen of the mixture. The ability of facultative anaerobes to remove oxygen is a key factor for maintaining a proper anaerobic digestion process (Turovskiy and Mathai, 2006).

During acetogenesis, volatile fatty acids are broken down to form acetate, H_2, and CO_2. The main product of acetogenesis is acetate; approximately 70% of methanogens utilize the acetoclastic pathway to form methane (Gray, 2004). Since acetogenesis and methanogenesis occur simultaneously during the digestion process, the products of acetogenesis are converted to biogas at the same rate at which they are formed. This is the reason for low levels of acetic acid and hydrogen in a properly operating digester. Given that acid forming bacteria have doubling times measured in hours while methane forming bacteria double at rates measured in days, the balance between acetogenesis and methanogenesis stages is critical to maintain proper digestion (Gray, 2004). Because of the high growth rate, acetogenic bacteria are tolerant to changes in pH and temperature (Turovskiy and Mathai, 2006). The kinetics of acetogenesis depends partly upon the growth rate expressed by methanogenic bacteria.

During the final stage of the anaerobic sludge digestion, i.e., methanogenesis, gasification process is needed to convert the organic acids to methane and carbon dioxide, i.e., biogas (Andreoli, 2007). Acid-splitting methane-forming bacteria are oxygen-intolerant, obligatory anaerobic archaea bacteria and are very sensitive to environmental conditions of temperature, pH, and anaerobiosis (Hammer and Hammer Jr., 2008). Internal buffering system helps methanogens to overcome changes in pH. This buffering capability prevents the accumulation of volatile fatty acids (Gray, 2004). The buffering system is based on bicarbonate in the liquid phase and carbon dioxide in the gas phase (Appels et al., 2008). The generation time of methane-forming bacteria ranges from a few days to several weeks under optimal conditions. It means that extended retention time in reactor is needed accordingly (Gerardi, 2006). A methanogenic anaerobic ecosystem is stabilized by its large unused capacity of hydrogen consumption which is buffering the partial pressure of dissolved hydrogen in the system at sufficiently low values to permit rapid fatty acid oxidation (Kaspar et al., 1978).

One of the most important factors affecting biological growth and activity is temperature. Mesophilic bacteria grow in a temperature range of 10 to 40°C, with an optimum of 37°C. Thermophilic bacteria have a range of approximately 45 to 75°C, with an optimum near 55°C (Hammer and Hammer Jr. 2008). Siegrist et al. presented a mathematical model to describe the dynamic behavior of mesophilic and thermophilic digestion with emphasis on acetotrophic methanogeneis and propionate degradation, as the steps that determine the anaerobic digestion process stability and the hydrolysis rate, which determines the degradation efficiency of particulate degradable organic carbon (Siegrist et al., 2002).

One of the important advantages of employing anaerobic digester for sludge stabilization is the generation of biogas which can be used for various applications. However, the heating value of the pure methane gas is higher than the biogas. Due to lower temperature and wastewater characteristics, anaerobic treatment processes are used in several European countries for proper industrial wastewater treatment. Table 1 shows the estimated quantity of biogas produced in various wastewater treatment processes by means of different approaches. The quantities of produced biogas related to the physical treatment, i.e., primary clarification are listed in Table 2.

Luetzer and Tietz have investigated the biogas production process under mesophilic and thermophilic conditions comprehensively (Luetzer and Tietze 2004).

Roediger et al. (1999) determined the quantity of produced biogas equals 700 to 800 Liter per kilogram decompsed organic solid waste. The resulted cumulative quantities of the produced biogas related to the mentioned investigations of Luetzer and Tietz under thermophilic and mesophilic conditions are shown in the figures 1 and 2.

In addition, under thermopilic operational condition, higher concentrations of NH_4-N, COD and P are available in digester for microorganisms. However, proper sludge dewatering capacity results under mesophilic operation condition.

Table 1. The quantity of biogas production as a result of the wastewater treatment process (biological treatment) (Kapp, 1984)

Modes of operation	Influent flow	Expected quantity of the produced biogas		
		L/day.person		
	grBOD$_5$/day.person	min	max	average
Primary Treatment + Aeration Basin (Sludge age 8 days , Nitrification in summer, sometimes with limited De-nitrification)	35	16.5	25	20.7
Primary Treatment + Aeration Basin (Sludge age 15 days, Nitrification and Denitrification throughout the day)	35	14.5	22	18.3
Primary Treatment + Aeration Basin (Sludge age 15 days, low physical treatment for solids removal)	48	10.5	15.9	13.2
Aeration Basin (Sludge age 15 days)	60	6.2	9.4	7.8
Simultaneous sludge stabilization (Sludge age 25 days)	60	3.5	5.3	4.4

Table 2. Effect of physical treatment, i.e., primary clarification on produced biogas (Luetzer and Tietze., 2004)

Detention time in primary settling tank (hour)	Primary sludge (grCOD/day.person)	The quantity of produced biogas (65% CH_4)	
		COD removal efficiency 55% (L/day.person)	COD removal efficiency 60% (L/day.person)
0.5	20	5.9	6.5
0.75	30	8.9	9.7
1	30	8.9	9.7
1.5	40	11.9	12.9
1.75	40	11.9	12.9

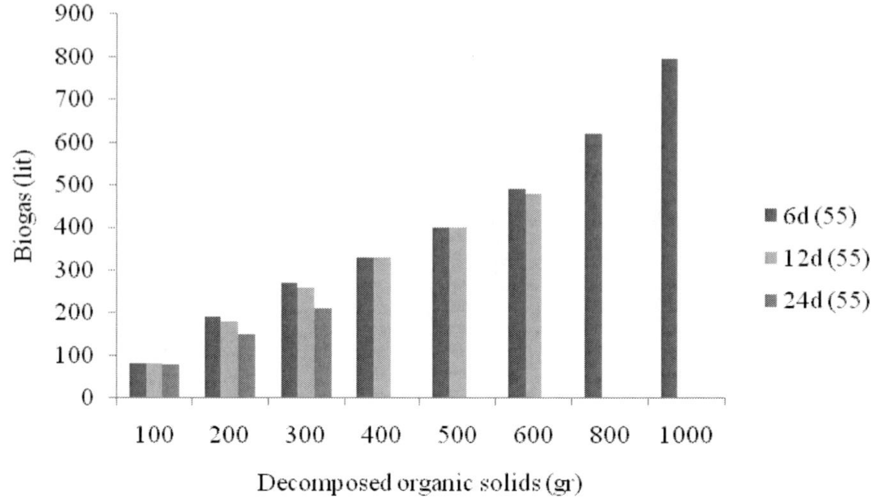

Figure 1. The quantities of biogas production resulted from organic solid waste decomposition under thermophilic condition. COD removal efficiency equals 55 percent, sludge ages equal 6, 12 and 24 days (Luetzer and Tietze 2004).

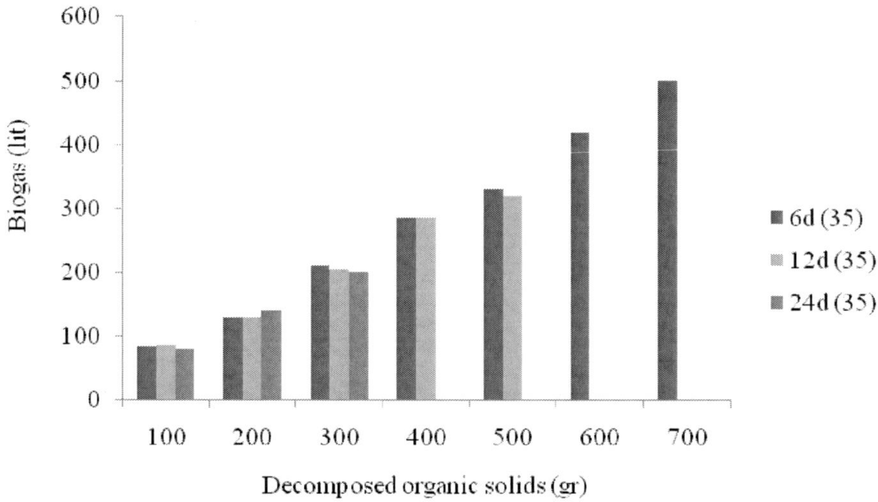

Figure 2. Cumulative quantities of biogas production resulted from organic solids decomposition in excess sludge stabilization process under mesophilic condition. COD removal efficiency equals 35 percent, sludge ages equal 6, 12 and 24 days (Luetzer and Tietze 2004).

Consequently, compared to mesophilic operated digesters, the thermophilic sludge stabilization process has following advantages:

- higher biogas production,
- higher decomposition of the organic solids.

2. Affecting Parameters

Significant abiotic factors in the anaerobic digester include temperature, pH, moisture content, substrate/carbon source, hydraulic retention time (HRT), ammonia-nitrogen, volatile fatty acids, total solids, volatile solids, and alkalinity. Table 3 lists the general operating conditions considering these parameters.

3. Inhibition

The effects of inhibitory substances on the anaerobic digestion process have been widely investigated over several decades. A wide variety of inhibitors are the primary cause of anaerobic digester upset or failure. Table 4 lists information on the inhibitory substances accordingly.

4. Pretreatment

The common preferred method for the sludge stabilization is the anaerobic digestion. However, this process is not optimized due to the hydrolysis step's limitation means low yield, high retention time, and high digester volume. Various pretreatment methods of waste activated sludge results in an improved sludge digestion efficiency. The existing methods include physical, physic-chemical, chemical, biological, thermal, as well as other methods such as ultrasonic pretreatment and so forth. Tables 5 to 8 list the classified common alternatives (Perez Elvira 2011; Tiehm et al., 2001; Weemaes and Verstraete 1998).

Table 3. Operating conditions for anaerobic sludge digestion

Factor	General operating range	Effect	References
Temperature	Mesophilic and Thermophilic conditions	- The increase of temperature causes rising reaction rate - The decrease of temperature reduces microbial growth, substrate utilization and biogas production. - The maximum temperature of the thermophilic digestion process should be less than 65 °C to avoid the adverse effect on the microbial population. • -The quantity of biogas production under thermophilic condition is higher than under mesophilic condition. • - The degradation rate of organic biomass under thermophilic condition is higher than mesophilic condition.	El-Mashad et al. 2004 Dela-Rubia et al., 2002 Bouallagui et al., 2009 Riau et al., 2010 Kim et al., 2006 Trzcinski and Stuckey, 2010 Zhu et al., 2009
pH	6.3 to 7.8	-Reduced quantity of the biogas production in case of inappropriate pH values. - The optimum pH value for biogas production equals 7. -The optimum range of pH value is 6.8 to 7.2.	El-Mashad et al., 2004 Huber et al., 1982 Yang and Okos., 1987
moisture	60 to 80 percent	- High humidity facilitate digestion process. -High methane production rate is obtainable in the mentioned range.	Hernandez-Berriel et al., 2008 Bouallagui et al., 2003
Substrate/ carbon source	……..	- Digestion rate depends on the type and structure of the substrate. -There are different carbon sources for various bacteria. -Carbohydrates are the most important substrate for the biogas production process occurrence.	Ghaniyari-Benis et al., 2009 Zhao et al., 2010 Dong et al., 2009

Factor	General operating range	Effect	References
Hydraulic Retention Time (HRT)	………	An increase in the hydraulic retention time results in high quantity of biogas production.	Turovskiy and Mathai, 2006
Ammonia-nitrogen	Concentration less than 5500 mg/L	-High ammonia concentration is classified as toxic (toxicity to bacteria) and inhibits the methane production. - The concentration equals 6,000 mg inhibits completely the methane production.	Tada, et al., 2005 Sawayama et al., 2004
Volatile Fatty Acids (VFA)	200 mg/L acetic acid (HAc)	-Excessive formation of the VFA decreases the pH value and inhibit the digestion reaction and consequently, can reduce the methane production. -Buffering system capacity is important to prevent the increasing pH value. -Excessive production of VFA causes hydrogen production that has an adverse impact on methane production.	Carucci et al., 2005 Georgacakis et al., 1982 Hill and Bolte, 2000
Total solids and volatile solids	TS> 10% of the total input	- Impact on loading rate -Optimum range (TS= 5% of total volume)	Wilkie, 2003 Møller et al., 2004
Alkalinity	1000 to 5000 mg/L as CaCO3	- The prevention of a decrease in pH value because of the VFA production.	Grady et al., 1999

Table 4. Inhibitors of anaerobic digestion process

Inhibitory substance		Range	Effect	References
Ammonia		More than 4200 mg/L	- Causes a decrease in methane production. - Has Toxic effect on methane maker bacteria.	Jarrell et al., 1987 Koster and Lettinga, 1988
Sulfide		100 to 800 mg/L a possible inhibitory substance; more than 800 mg/L definitely a inhibitory substance	- Inhibits the methane formation - Toxicity to bacteria - The quantity of inhibitory sulfide depends on the pH value.	Harada et al., 1994 Colleran et al., 1998 Parkin et al., 1990 O'Flaherty et al., 1998
Light metals ions	Aluminum	More than 1000 mg/L	By adhering to the cell membrane can inhibit the growth of bacteria.	Cabirol et al., 2003
	Calcium	More than 7000 mg/L	- Effective material for the growth of methane-forming bacteria. - High calcium concentration causes corrosion of digestion facilities. - Causes nutrient removal and the loss of buffering system capacity - High calcium concentration causes reduced methane production.	Murray and Zinder, 1985 van Langerak et al., 1998 Jackson-Moss et al., 1989
	Magnesium	More than 400 mg/L	Inhibitory impact on the growth of some bacteria	Schmidt and Ahring, 1993
	Potassium	……	Inhibitory impact on the acidification step	Fernandez and Forster, 1994
	Sodium	More than 8000 mg/L	Causes a cellular metabolism disturbance	Mende´z et al., 1995 McCarty, 1964

Inhibitory substance	Range	Effect	References
Heavy metals	………..	- High concentration causes toxicity. - Inhibits the biological process in the digester.	Sterritt and Lester, 1980 Takashima and Speece, 1989
Organics	……..	- Organic compounds (benzene and halogenated benzene, phenol, alcohols, esters...) cause swelling and permeability of the membranes of microorganisms. - Cause failure of the digestion process.	Sikkema et al., 1994 Renard et al., 1993 van Beelen and van Vlaardingen, 1994 Dimirer and Speece, 1998 Hayward and Lau, 1989

Table 5. Physical pretreatment methods

	Processes	Advantages	Disadvantages	References
Physical pretreatment	Milling includes - Ball mill and cutting mill, - Hammer mill, - -Colloid mill.	- The increase of 19% in VS degradation, - An Improvement in hydrolysis occurrence by decreasing strength and size of solids, - - Increasing biogas production.	- High energy costs, - - Not able to remove the lignin as an inhibitor.	Mooney et al., 1999 Zhang et al., 1999 Baier and Schmidheiny, 1997
	Irradiation includes - Microwave - irradiation, - Gamma-ray - irradiation, - Electron-beam - irradiation.	- An Improvement in hydrolysis occurrence, - Decreasing polymerization degree, - - A combination of the process and chemical methods is possible.	- High cost, - Not able to - remove the lignin, - -Difficult to use.	Kumakura et al., 1984 Lafitte-Trouque et al., 2002 Eskicioglu et al., 2007

Table 6. Physico-chemical pretreatment methods

	Processes	Advantages	Disadvantages	References
Physico-chemical pretreatment	Explosion includes - Steam explosion - CO_2 explosion - Ammonia fiber explosion (AFEX)	- Increasing Biogas Production, -An Improvement in hydrolysis occurrence, -Applicable to wastewater treatment plants, - Carbon dioxide explosion is an environmental friendly and low cost process, - Availability of chemical materials in pretreatment process, - decreasing lignin quantity in the AFEX process,	- High energy consumption, - Increasing temperature and pressure , - Carbon dioxide explosion method is sensitive to temperature, - The ammonia reuse is in the AFEX process required for decreasing the cost and preventing from entering the environment.	Sun et al., 2005 Negro et al., 2003 Chum et al., 1985 Varga et al., 2004 Alizadeh et al.,2005 Eggeman et al., 2005 Zheng et al., 1996 Zheng et al., 1995

Table 7. Chemical pretreatment methods

	Processes	Advantages	Disadvantages	References
Chemical pretreatment	Alkaline hydrolysis	-Solubilising VS equals 40 to 50 percent, reduction in lignin quantity equals 40 percent, reduction in hemicellulose equals 30 percent, -The methane production quantity becomes twice.	The process occurrence requires a high retention time at low temperature.	Kassim et al., 1986 Xu et al., 2007 Zhao et al., 2007 Tanaka et al., 1997 Tanaka et al., 2002
	Alkaline peroxide	-Improving enzyme hydrolysis, -Sugar fermentation and transformation to sugar enzyme up to 97 percent, -Lignin reduction equals 80 to 90 percent.	………	Saha and Cotta., 2006 Saha and Cotta., 2007 Sun et al., 2005 Curreli et al., 1997
	Organosolv process	-Improving enzyme activity, -The availability of lignin to electricity and heat production.	-Requiring to Catalyst for temperature reduction -Requiring to solvents which must be removed after post-treatment	Pan et al., 2005 Sun et al., 2002

	Processes	Advantages	Disadvantages	References
	Wet oxidation	-Hemicellulose reduction, -Increasing quantity of the organic acid production, -Increasing quantity of the methane production equals 35 to 70 percent.	Temperature control is very important	Saha and Hemicellulose., 2003 Garrote et al., 1999 Galbe et al., 2002
	Ozonolysis	-Hemicellulose reduction, -Reduction of lignin quantity, -Improvement of the digestion process as a result of the reduction in toxic phenol, - Increasing quantity of the methane production.	High cost	Vidal et al., 1988 Sun et al., 2002 Goel et al., 2003
	Acid hydrolysis	-Improvement of the hydrolysis; Lignin quantities reduction, -Biogas production increase from 30 to 70 percent.	-The high acid concentration causes corrosion, -Requiring to neutralization, -Activation of the inhibitors at low pH.	Taherzadeh et al., 2007 Jones et al., 1984 Kivaisi et al., 1994

Table 8. Other pretreatment methods

	Processes	Advantages	Disadvantages	Reference
Other pretreatment methods	Thermal pretreatment	- -An increase in the VSS degradation equals 30 percent, -An increase in the - methane production equals 100 percent.	High energy consumption	Li and Noike, 1992
	Biological pretreatment	- -Enzyme hydrolysis improvement, - -An increase in the biogas production equals 86 percent under thermophilic condition.	…………	Sun et al., 2002 Gavala et al., 2003
	Ultrasonic pretreatment	- -Cell wall deterioration, - -Increased ratio of solubilisation up to 80 percent. - - An increase in the biogas production equals 50 percent.	High energy consumption	Shimizu et al., 1993 Appels et al., 2008

REFERENCES

Alizadeh, H., Teymouri, F., Gilbert, T.I. & Dale, B.E. (2005). Pretreatment of switchgrass by ammonia fiber explosion (AFEX). *Appl. Biochem. Biotechnol 124*, 1133-1141.

Appels, L., Baeyens, J., Degrève, J. & Dewil, R. (2008). Principles and potential of the anaerobic digestion of waste-activated sludge. *Progress in Energy and Combustion Science 34*, 755-781.

Appels, L., Dewil, R. & Baeyens, J. (2008). Ultrasonically enhanced anaerobic digestion ofwaste activated sludge. *Int. J. Sustain. Eng, 1.*

Baier, U., Schmidheiny, P. (1997). Enhances anaerobic degradation of mechanically disintegrated sludge. *Water Sci. Technol. 37,* 137–143.

Borja, R., Banks, C.J., Martín, A. & Khalfaoui, B. (1995). Anaerobic digestion of palm oil mill effluent and condensation water waste: an overall kinetic model for methane production and substrate. *Bioprocess and Biosystems Engineering 13*, 87–95.

Bouallagui, H., Cheikh, R.B., Marouani, L. & Hamdi, M. (2003). Mesophilic biogas production from fruit and vegetable waste in tubular digester. *Bioresour. Technol 86*, 85–89.

Bouallagui, H., Rachdi, B., Gannoun, H. & Hamdi, M. (2009). Mesophilic and thermophilic anaerobic co-digestion of abattoir wastewater and fruit and vegetable waste in anaerobic sequencing batch reactors. *Biodegradation 20*, 401–409.

Cabirol, N., Barraga´n, E.J., Dura´n, A. & Noyola, A. (2003). Effect of aluminum and sulphate on anaerobic digestion of sludge from wastewater enhanced primary treatment. *Water Sci. Technol 48 (6)*, 235–240.

Carucci, G., Carrasco, F., Trifoni, K., Majone, M. & Beccari, M. (2005). Anaerobic Digestion of Food Industry Wastes: Effect of Codigestion on Methane Yield. *J. Environ. Engineering, ASCE,* 1037.

Chen, Y., Cheng, J.J. & Creamer, K.S. (2007). Inhibition of anaerobic digestion process: A review, *Bioresource Technol. 99*, 4044-4064.

Chum, H.L., Douglas, L.J., Feinberg, D.A. & Schroeder, H.A. (1985). Evaluation of pretreatments of biomass for enzymatic hydrolysis of cellulose. *Solar Energy Research Institute: Golden,* Colorado, 1-64.

Colleran, E., Pender, S., Phipott, U., O'Flaherty, V. & Leahy, B. (1998). Full-scale and laboratory-scale anaerobic treatment of citric acid production wastewater. *Biodegradation 9*, 233–245.

Curreli, N., Fadda, M.B., Rescigno, A., Rinaldi, A.C., Soddu, G., Sollai, F., Vaccargiu, S., Sanjust, E. & Rinaldi, A. (1997). Mild alkaline/oxidative pretreatment of wheat straw. *Process Biochem. 32*, 665-670.

Dela-Rubia, M.A., Perez, M., Romero, L.I. & Sales, D. (2002). Anaerobic mesophilic and thermophilic municipal sludge digestion. *Chem. Biochem. Eng. Qual 16*, 119–124.

Dimirer, G.N. Speece, R.E. (1998). Biotransformation of four 3-carbon compounds (acrolein, acrylic acid, allyl alcohol and N-propanol) in UASB reactors. *Water Res. 32 (3)*, 747–759.

Dong, L., Zhenhong, Y., Yongming, S., Xiaoying, K. & Yu, Z. (2009). Hydrogen production characteristics of organic fraction of municipal solid wastes by anaerobic mixed culture fermentation. *Int. J. Hydr. Energy 34*, 812–820.

Eggeman, T., Elander, R.T. (2005). Process and economic analysis of pretreatment technologies. *Bioresource Technol. 96*, 2019-2025.

El-Mashad, H.M., Zeeman, G., van Loon, W.K.P., Bot Gerard, P.A. & Lettinga, G. (2004). Effect of Temperature and Temperature Fluctuation on Thermophilic Anaerobic Digestion of Cattle Manure. *Bioresource Technol. 95*, 191.

Eskicioglu, C., Terzian, N., Kennedy, K.J., Droste, R.L. & Hamoda, M. (2007). Athermal microwaveeffects for enhancing digestibility of waste activated sludge. *Water Res. 41*, 2457-2466.

Fernandez, N., Forster, C.F. (1994). The anaerobic digestion of simulated coffee waste using thermophilic and mesophilic upflow filters. *Process Safety Environ. Protect. 72 (B1)*, 15–20.

Fogarty, W.M., Kelly, C.T. (1979). Developments in microbial extracellular enzymes. *Topics in Enzyme and Fermentation Biotechnology III, Ellis Horwood,* Chichester, 45.

Galbe, M., Zacchi, G. (2002). A review of the production of ethanol from softwood. *Appl. Microbiol. Biotechnol. 59*, 618-628.

Garrote, G., Dominguez, H. & Parajo, J.C. (1999).Hydrothermal processing of lignocellulosic materials. *Holz als Roh-und Werkstoff 57*, 191-202.

Gavala, H.N., Yenal, U., Skiadas, I.V., Westermann, P. & Ahring, B.K. (2003). Mesophilic and thermophilic anaerobic digestion of primary and secondary sludge. Effect of pre-treatment at elevated temperature, *Water Res. 37 (19)*, 4561–4572.

Ghaniyari-Benis, S., Borja, R., Ali Monemian, S. & Goodarzi, V. (2009). Anaerobic treatment of synthetic medium-strength wastewater using a multistage biofilm reactor. *Bioresource Technol. 100*, 1740–1745.

Goel, R., Tokutomi, T., Yasui, H. & Noike, T. (2003). Optimal process configuration for anaerobic digestion with ozonation. *Water Sci. Technol. 48*, 85-96.

Grady, J.C.P.L., Daigger, G.T. & Lim, H.C. (1999). Biological Wastewater Treatment, 2nd ed. Marcel Dekker, Inc.: 599-600.

Gray, N.F. (2004). Biology of Wastewater Treatment, 2nd ed. Imperial College Press, 146: 262-270, 735-757.

Hammer, M.J., Hammer, M.J. Jr. (2008). Water and Wastewater Technology. Sixth Ed., Pearson, Prentice Hall, New Jersey, Columbus, Ohio.

Harada, H., Uemura, S. & Monomoi, K. (1994). Interactions between sulphate-reducing bacteria and methane producing bacteria in UASB reactors fed with low strength wastes containing different levels of sulphate. *Water Res.,* 355–367.

Hayward, G., Lau, I. (1989). Toxicity of organic solvents to fatty acid forming bacteria. *Can. J. Chem. Eng. 67*, 157–161.

Hernandez-Berriel, M.C., Benavides, L.M., Perez, D.J.G. & Delgado, O.B. (2008). The effect of moisture regimes on the anaerobic degradation of municipal solid waste from Metepec (Mexico). *Waste Managem. 28*, 14–20.

Hill, D.T., Bolte, J.P. (2000). Methane production from the Low Solid Concentration Liquid Swine Waste Using Conventional Anaerobic Fermentation. *Bioresource Technol. 74*, 241.

Huber, H., Thomm, M., Konig, H., Thies, G. & Stetter, K.O. (1982). Methanococeus thermolithotrophicus, a novel thermophilic lithotrophic methanogen. *Arch. Microbiol. 132*, 47–50.

Jackson-Moss, C.A., Duncan, J.R. & Cooper, D.R. (1989). The effect of calcium on anaerobic digestion. *Biotechnol. Lett. 11 (3)*, 219–224.

Jarrell, K.F., Saulnier, M. & Ley, A. (1987). Inhibition of methanogenesis in pure cultures by ammonia, fatty acids, and heavy metals, and protection against heavy metal toxicity by sewage sludge. *Can. J. Microbiol. 33*, 551–555.

Jones, J., Semrau, K. (1984). Wood hydrolysis for ethanol production previous experience and the economics of selected processes. *Biomass, 5*, 109-135.

Kasper, H.F., Wuhrmann, K. (1978). Kinetic Parameters and Relative Turnovers of Some Important Catabolic Reactions in Digesting Sludge. *American Society of Microbiology, Applied and Environmental Microbiology 36(1)*,1-7.

Kassim, E.A., El-Shahed, A.S. (1986). Enzymatic and chemical hydrolysis of certain cellulosic materials. *Agr. Wastes 17*, 229-233.

Kim, J.K., Oh, B.R., Chun, Y.N. & Kim, S.W. (2006). Effects of temperature and hydraulic retention time on anaerobic digestion of food waste. *J. Biosci. Bioeng. 102*, 328– 332.

Kim, T.H., Lee, Y.Y. (2006). Fractionation of corn stover by hot-water and aqueous ammonia treatment. *Bioresource Technol. 97*, 224-232.

Kivaisi, A.K., Eliapenda, S. (1994). Pretreatment of bagasse and coconut fibres for enhanced anaerobic degradation by rumen microorganisms. *Renew. Energy 5*, 791-795.

Koster, I.W., Lettinga, G. (1988). Anaerobic digestion at extreme ammonia concentrations. *Biol. Wastes 25*, 51–59.

Kumakura, M., Kaetsu, I. (1984). Pretreatment by radiation and acids of chaff and its effect on enzymatic hydrolysis of cellulose. *Agr. Wastes 9*, 279-287.

Lafitte-Trouque, S., Forster, C.F. (2002). The use of ultrasound and gamma-irradiation as pre-treatments for the anaerobic digestion of waste activated sludge at mesophilic and thermophilic temperatures. *Bioresource Technol. 84*, 113-118.

McCarty, P.L. (1964). Anaerobic waste treatment fundamentals. *Public Works 95(9)*, 107–112.

Mende´z, R., Lema, J.M. & Soto, M. (1995). Treatment of seafood-processing wastewaters in mesophilic and thermophilic anaerobic filters. *Water Environ. Res. 67 (1)*, 33-45.

Mooney, C.A., Mansfield, S.D. Beatson, R.P. & Saddler, J.N. (1999). The effect of fiber characteristics on hydrolysis and cellulase accessibility to softwood substrates. *Enzyme Microb. Tech. 25*, 644-650.

Murray, P.A., Zinder, Z.H. (1985). Nutritional requirements of Methanosarcina sp. strain TM-1. *Appl. Environ. Microbiol. 50*, 49-56.

Negro, M.J., Manzanares, P., Ballesteros, I., Oliva, J.M., Cabanas, A. & Ballesteros, M. (2003). Hydrothermal pretreatment conditions to enhance ethanol production from poplar biomass. *Appl. Biochem. Biotechnol. 105*, 87-100.

O'Flaherty, V., Mahony, T., O'Kennedy, R. & Colleran, E. (1998). Effect of pH on growth kinetics and sulphide toxicity thresholds of a range of methanogenic, syntrophic and sulphate-reducing bacteria. *Process Biochem. 33 (5)*, 555–569.

Pan, X., Arato, C., Gilkes, N., Gregg, D., Mabee, W., Pye, K., Xiao, Z., Zhang, X. & Saddler, J. (2005). Biorefining of softwoods using ethanol organosolv pulping: Preliminary evaluation of process streams for manufacture of fuel-grade ethanol and co-products. *Biotechnol.Bioeng. 90*, 473-481.

Parkin, G.F., Lynch, N.A., Kuo, W., van Keuren, E.L. & Bhattacharya, S.K. (1990). Interaction between sulfate reducers and methanogens fed acetate and propionate. *Res. J. Water Pollut. Control Fed. 62*, 780–788.

Perez Elvira, S.I. (2011). Advanced Anaerobic Digestion-Evaluating the Options. *Tecnho Focus, Asian Water,* 17-22.

Renard, P., Bouillon, C., Neveau, H. & Nyns, E. (1993). Toxicity of a mixture of polychlorinated organic compounds towards an unacclimated methanogenic consortium. *Biotechnol. Lett. 15 (2)*,195–200.

Riau, V., De la Rubia, M.A. & Pérez, M. (2010). Temperature-phased anaerobic digestion (TPAD) to obtain class A biosolids: a semi-continuous study. *Bioresource Technol. 101*, 2706–2712.

Saha, B.C. (2003). Hemicellulose bioconversion.Ind. *Microbiol Biotechnol. 30*, 279-291.

Saha, B.C., Cotta, M.A. (2006). Ethanol production from alkaline peroxide pretreated enzymatically saccharified wheat straw. *Biotechnol.Progr. 22*, 449-453.

Saha, B.C., Cotta, M.A. (2007). Enzymatic saccharification and fermentation of alkaline peroxide pretreated rice hulls to ethanol. *Enzyme Microb. Technol. 41*, 528-532.

Sawayama, S., Tada, C., Tsukahara, K. & Yagishita, T. (2004). Effect of ammonium addition on methanogenic community in a fluidized bed anaerobic digestion. *J. Biosci. Bioen. 97*, 65–70.

Schmidt, J.E., Ahring, B.K. (1993). Effects of magnesium on thermophilic acetate-degrading granules in upflow anaerobic sludge blanket (UASB) reactors. *Enzyme Microbiol. Technol. 15*, 304–310.

Speece, R.E., Parkin, G.F. (1983). The response of methane bacteria to toxicity. In: *Proceedings of the 3rd International Symposium on Anaerobic Digestion*, Boston, MA.

Sterritt, R.M., Lester, J.N. (1980). Interaction of heavy metals with bacteria. *Sci. Total Environ 14 (1)*, 517.

Shimizu, T., Kudo, K. & Nasu, Y. (1993). Anaerobic waste-activated sludge digestion, a bioconversion mechanism and kinetic model. *Biotechnol. Bioeng. 41*, 1082–1091.

Siegrist, H., Vogt, D., Garcia-Heras, J.L. & Gujer, W. (2002). Mathematical Model for Meso- and Thermophilic Anaerobic Sewage Sludge Digestion, *Environmental Science and Technology, 36,* 1113-1123.

Sikkema, J., De Bont, J.A.M. & Poolman, B. (1994). Interactions of cyclic hydrocarbons with biological membranes, *Journal of Biol. Chemistry, 26*, 8022–8028.

Stronach, S.M., Rudd, T. & Lester, J.N. (1986). Anaerobic Digestion Processes in Industrial Wastewater Treatment, Springer: 16–36.

Sun,X.F., Xu, F., Sun, R.C., Fowler, P. & Bairdd, M.S. (2005). Characteristics of degraded cellulose obtained from steam-exploded wheat straw. *Carbohyd. Res. 340*, 97-106.

Sun, Y., Cheng J. (2002). Hydrolysis of lignocellulosic materials for ethanol production: A review. *Bioresource Technol. 83*, 1-11.

Sun, Y., Cheng, J. (2005). Dilute acid pretreatment of rye straw and bermudagrass for ethanol production. *Bioresource Technol. 96*, 1599-1606.

Taherzadeh, M.J., Karimi, K. (2007). Acid-based hydrolysis processes for ethanol from lignocellulosic materials: A review. *Bioresources 2*, 472-499.

Tanaka, S., Kamiyama, K. (2002). Thermochemical pretreatment in the anaerobic digestion of waste activated sludge. *Water Sci. Technol. 46*, 173-179.

Tanaka, S., Kobayashi, T., Kamiyama, K. & Bildan, M. (1997). Effects of thermo- chemical pretreatment on the anaerobic digestion of waste activated sludge. *Water Sci. Technol. 35*, 209-215.

Tiehm, A., Nickel, K., Zellhorn, M. & Neis, U. (2001). Ultrasonic waste activated sludge disintegration for improving anaerobic stabilization. *Water Res. 35*, 2003–2009.

Trzcinski, A.P., Stuckey, D.C. (2010). Treatment of municipal solid waste leachate using a submerged anaerobic membrane bioreactor at mesophilic and psychrophilic temperatures: analysis of recalcitrants in the permeate using GC-MS. *Water Res. 44*, 671–680.

Turovskiy, I.S., Mathai, P.K. (2006). Wastewater Sludge Processing. John Wiley and Sons, Inc.

van Beelen, P., van Vlaardingen, P.V. (1994). Toxic effects of pollutants on the mineralization of 4 chlorophenol and benzoate in methanogenic river sediment. *Environ. Toxicol. Chem. 13 (7)*, 1051–1060.

van Haandel, A.C., Letinga, G. (1994). Anaerobic Sewage Treatment: A Practical Guide for Regions with a Hot Climate, John Wiley and Sons, Inc., 34-35.

van Langerak, E.P.A., Gonzales-Gil, G., van Aelst, A., van Lier, J.B., Hamelers, H.V.M. & Lettinga, G. (1998). Effects of high calcium concentrations on the development of methanogenic sludge in upflow anaerobic sludge bed (UASB) reactors. *Water Res. 32 (4)*, 1255–1263.

Varga, E., Reczey, K. & Zacchi, G. (2004). Optimization of steam pretreatment of corn stover to enhance enzymatic digestibility. *Appl. Biochem. Biotechnol. 113*, 509-523.

Vidal, P.F., Molinier, J. (1988). Ozonolysis of Lignin – Improvement of in vitro digestibility of poplar sawdust. *Biomass 16*, 1-17.

Weemaes, M.P.J., Verstraete, W. (1998). Evaluation of current wet sludge disintegration techniques. *J. Chem. Technol. Biotechnol. 73*, 83–92.

Wilkie, A.C. (2003). Anaerobic Digestion of Flushed Dairy Manure. Proceedings – Anaerobic Digester Technology Applications in Animal Agriculture – A National Summit, from *the Water Environment Federation,<http://www.epa.gov/agstar/pdf/wefjune2003.pdf>* Accessed Nov. 2005.

Yang, S.T., Okos, (1987). Kinetic study and mathematical modeling of methanogenesis of acetate using pure cultures of methanogens. *Biotechnol.Bioeng. 30*, 661–667.

Zhang, R.H., Zhang, Z.Q. (1999). Biogasification of rice straw with an anaerobic-phased solids digester system. *Bioresource Technol. 68*, 235-245.

Zhao, X., Zhang, L. & Liu, D. (2007). Comparative study on chemical pretreatment methods for improving enzymatic digestibility of crofton weed stem. *Bioresource Technol. 99*, 3729-3736.

Zhao, Y., Wang, A. & Ren, N. (2010). Effect of carbon sources on sulfidogenic bacterial communities during the starting-up of acidogenic sulfate-reducing bioreactors. *Bioresource Technol. 101*, 2952–2959.

Zheng, Y., Lin, H.M., Wen, J., Cao, N., Yu, X. & Tsao, G.T. (1995). Supercritical carbon dioxide explosion as a pretreatment for cellulose hydrolysis. *Biotechnol. Lett. 17*, 845-850.

Zheng, Y., Tsao, G.T. (1996). Avicel hydrolysis by cellulase enzyme in supercritical CO2. *Biotechnol. Lett. 18*, 451-454.

Zhu, B., Gikas, P., Zhang, R., Lord, J., Jenkins, B. & Li, X. (2009). Characteristics and biogas production potential of municipal solid wastes pretreated with a rotary drum reactor. *Bioresour. Technol. 100*, 1122–1129.

In: Methanogenesis
Editor: Gagik Badalians Gholikandi
ISBN: 978-1-63321-567-2

Chapter 7

BIOENERGY GENERATION: METHANOGENESIS IN BIO-ELECTROCHEMICAL SYSTEMS

Zahra Tizmaghz, Amirreza Arashi and Gagik Badalians Gholikandi*
Faculty of Water Engineering and Environment,
Shahid Beheshti University, A. C., Tehran, Iran

ABSTRACT

Bioelectrochemical systems are capable to treat organic waste solids and wastewater and generate electrical energy simultaneously. These technologies present many exciting opportunities and chronologically contain different challenges to be utilized in various applications. There are several types of bioelectrochemical systems regarding the purpose of the applications, divided into two major groups, i.e., electricity generation and valuable chemicals production employing microbial cells. Methane is commonly observed in bioelectrochemical systems in the course of hydrogen production process or inside of the existing biofilms in the presence of methanogens. These microorganisms are able to produce methane gas utilizing electrons directly obtained from currents. Investigating the fundamental mechanisms of this process is prerequisite

* Corresponding author: Hakimieh- Tehranpars, Shahid Abbaspour Blvd., PO Code 16589-53571, PO Box 16765-1719, Tehran, Iran. Email: g.badalians@yahoo.com.

of generating methane gas by electrical current in different operation conditions and higher efficiency. The methane formation in these systems is dependent on available substrate, reactor configuration and the cells condition. In this chapter bioenergy generation in bioelectrochemical systems has been discussed as a novel technology. The related technologies have been divided into four groups and mechanisms, and the advantages and limitations of each one have been described. Consequently, the conceivability and the effect of the methanogenesis process on their behavior have been investigated.

Keywords: Bioenergy, methanogenesis, bioelectrochemical system, energy generation

1. Bioelectrochemical Systems

In recent years, due to the necessity of decreasing fossil fuels consumption, researches on sustainable energy sources have been increased noticeably. (Logan et al., 2006; Logan et al., 2008). Bioelectrochemical systems are able to simultaneously treat wastewater and generate electrical energy (Chaudhuri and Lovley 2003; Aelterman et al., 2006; Jacobson et al., 2011; Logan et al., 2006; Lovley 1988; Zhang and Angelidaki 2012a; Zhang and Angelidaki 2013). Two new methods of bioenergy generation from biomass include electricity production employing microbial fuel cells (MFCs) and hydrogen production by electrohydrogenesis utilizing microbial electrolysis cells (MECs) (Logan et al., 2006; Logan et al., 2008). These technologies present exciting opportunities and face different challenges in order to be utilized in various applications other than energy production such as wastewater treatment; on the other hand the costs of these technologies are relatively high. It seems that increasing the costs of generating electrical power as well as encouragements to carbon-neutral processes attract more attentions and sufficient investments, in the near future (Logan and Rabaey 2012).

Aerobic wastewater treatment processes containing high loads of organic materials, consume noticeable amounts of energy. For example in the United States, is approximately 3% of the whole produced energy (Bae et al., 2011). On the other hand, wastewater (mixture of industrial, domestic and agricultural wastewater) averagely contains about 17 GW power (Logan and Rabaey 2012). It means that by capturing and utilizing this potential energy as a new source of electrical power, the required energy for wastewater treatment process is obtainable by itself. Moreover, by efficient management of agricultural wastes as organic biomass, larger amounts of energy can be produced. Therefore, related cost-effective methods are supposed to be developed in order to employ this potential energy. During capturing process of the biomass potential energy, some useful materials such as biofuel or frequently used chemicals can be generated by means of bioelectrochemical methods (Logan and Rabaey 2012).

The capturing processes in a bioelectrochemical system (microbial electrolysis cell) are handled by several electrochemical reactions through microorganisms. The reactions take place on the electrodes by planktonic or biofilm cells. The anodic reactions produce current and cathodic reactions consume the current or electrons. Also at the anode, polymeric materials degraded to simple molecules, such as fatty acids and hydrogen, which can be used for current generation consequently. Other bioconversion reactions, such as conversion of glycerol to ethanol, can produce compounds for current generation, too. At the cathode useful chemicals are produced which can be utilized in chemical industries (e.g., hydrogen peroxide) or energy generation (such as acetate in the anodic reactions)

1.1. Microbial Fuel Cell (MFC)

The concept of the catalyzing ability of microorganisms in fuel cells was explored in the 1970s (Roller et al., 1984) and the application of MFCs for domestic wastewater treatment was first presented in 1991 (Habermann and Pommer 1991). In recent years, the MFCs with the capability of providing considerable opportunities for practical applications have been developed with an enhanced power output (Liu and Logan 2004; Liu et al., 2004; Rabaey et al., 2003; Rabaey et al., 2004).

MFC–based technologies are promising methods to produce direct electrical power from waste biomass, wastewater treatment, and capable of saving salinity gradient in salt- and fresh water sources in the form of electrical energy (Liu et al., 2004; Cusick et al., 2010). MFC is a bioelectrochemical system that can be employed simultaneously to produce electrical power and treat wastewater. The MFCs do not require metal catalysts at the anode. Instead, they employ microorganisms which are able to biologically oxidize organic matter and transfer electrons to the anode. In an air cathode MFC, exo-electrogenic bacteria oxidize substrate, and then transfer electrons through the anode which causes the reduction of oxygen at the air cathode. It means that this method is applicable to wastewater treatment. In other words, the dissolved organic materials can be oxidized by exo-electrogenic microorganisms and electrons will be released to an electrode. Almost any source of biodegradable organic matter, including carbohydrates and proteins can be used in a MFC to generate power. The flexibility of microorganisms to utilize a wide range of organic materials makes the MFC a promising technology for bioelectricity generation from biomass. Therefore, MFCs have been used to convert the available organic matter of wastewater (Ahn and Logan 2009; Logan et al., 2008; Patil et al., 2009; Wen et al., 2009), aquatic sediments (Miller and Oremland 2008), and associated with plants (DeSchamphelaire et al., 2008) into electrical current (Kiely et al., 2011). The produced electrons transfer to the anode of a MFC can be performed either through direct contact (via highly conductive nanowires or membrane-associated proteins) (Lovley 2008) or via soluble electron shuttles (Rabaey et al., 2005).

In contrast to anaerobic digestion, a MFC generates electrical current and a gas containing mainly carbon dioxide (CO_2). MFCs have operational and functional advantages over the current technologies employed to generate energy from organic matter. These advantages are:

1. Converting substrate potential energy directly to electricity: enables high conversion efficiency.
2. Operating at ambient, and even at low temperatures: make them individual among current bioenergy processes.
3. Not requiring gas treatment: the produced gases are enriched in CO_2 and do not have useful energy content.
4. Requiring no energy input for aeration: the cathode is passively aerated.
5. Applicable for widespread applications: in locations lacking electrical infrastructures or to expand the diversity of fuels to satisfy energy requirements.

Decreasing the cost of the materials used in MFCs is crucial for practical applications. The oxidation of the fuel at the anode in a MFC is not a true catalysis step, as the microorganisms derive energy from the substrate oxidation. In the cathode, the reduction of oxygen is usually catalyzed by a costly metal catalyst, such as platinum. Depending on the energy released by the bacteria and energy lost at the cathode, the total produced energy is resulted. The cathode accounts for the main part of the total capital cost (Rozendal et al., 2008) and its surface area and materials generally limit higher power generation in MFCs (Cheng et al., 2006). Therefore, finding low-cost materials and efficient cathode structural design are important for improvement in cost effectiveness and performance of MFCs. Several studies have been focused on finding reasonable materials for cathode recent years. Carbon based materials are widely used as the electrode for air cathode MFCs. The high cost of some of these materials, such as carbon cloth inhibits their utilization for large scale systems. Low cost cathodes, have been made by applying conductive coatings on inexpensive membranes have relatively low power densities (Zuo et al., 2007).

In order to evaluate the metabolism of the bacteria and bacterial electricity generation, metabolic pathways that create microbial electron and proton flows must be determined. Moreover, the type of the available substrate (Rabaey et al., 2004; Kim et al., 2004) and the potential of the anode are impressive affecting parameters in this regard. By increasing the MFC's current, the potential of the anode decreases and accordingly forces the bacteria to deliver the electrons via more-reduced complexes. Therefore, the anode's potential determines the redox potential of the final bacterial electron shuttle, and also, the metabolism.

According to the anode potential, several different metabolism routes can be introduced as follows (Rabaey et al., 2004; Kim et al., 2004):

- high redox oxidative metabolism;
- medium to low redox oxidative metabolism; and
- fermentation

The organisms reported to date in MFCs vary from aerobes and facultative anaerobes to strict anaerobes (Zuo et al., 2007). Figure 1 demonstrates a schematic view to MFC and the related parameters in each part.

According to Figure 1, the complex environment of the MFC has many opportunities for scientific study and contains affecting challenges for commercial applications. An electrogenic biofilm forms on the anode, and its development is dependent on the specific electrode materials and properties. Inert materials of biofilm such as solids or dead cells can limit the cells access to the electrode. Electroactive cells can be in direct or indirect contact with the electrode and competing cells like methanogens which do not require this contact and have a competitive advantage outside of the biofilm. The pH gradient between the biofilm and bulk electrolyte leads to loss of electrochemical potential.

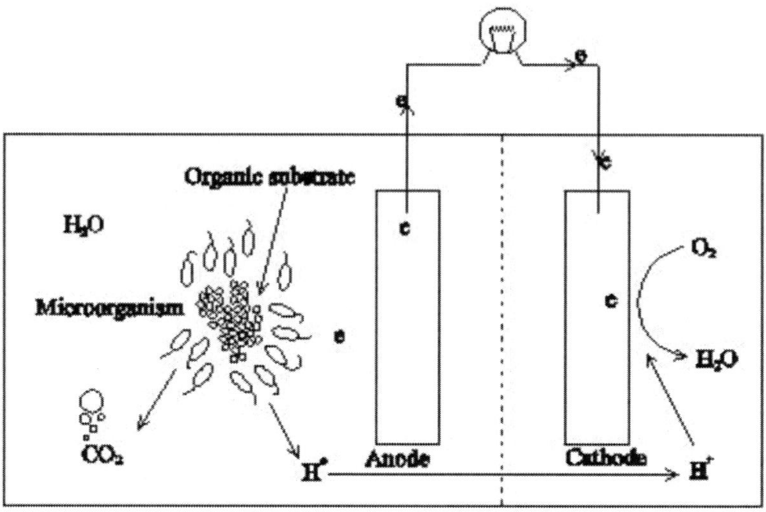

Figure 1. A schematic view of MFC structure

Solids in the wastewater can block the electrodes and affect its availability. The generated protons at anode and hydroxyl ions at cathode should be transported to the counter-electrode in competment to other ions like chloride and sodium and change the bulk pH, which in turn affects the process. The cathode catalyzes oxygen reduction. However, it can be fouled accompanied by bacteria or organic and inorganic matters, leading to the reduction in catalysis performance. Therefore a separator layer should be employed (Kim et al., 2004). The formed biofilm on the anode electrode is suitable for methanogens in order to produce methane using organic material. Amount of produced methane gas within the biofilm is negligible because of restrictive anaerobic conditions.

1.2. Microbial Electrolysis Cell (MEC)

Microbial electrolysis cell (MEC) has recently become known as a novel process for the renewable and sustainable production of biofuels or valuable chemicals from biomass (Clauwaert and Verstraete 2009, Wagner et al., 2009; Hamelers et al., 2010).

MEC as one of the typical Bioelectrochemical systems had attracted attention at the early stage of Bioelectrochemical systems research (Cheng et al., 2006; He and Mansfeld, 2009; Liu et al., 2005a; Logan 2005; Rabaey et al., 2005). MEC process has several advantages over other biological hydrogen production processes. Various biomasses such as cellulose, glucose, glycerol, acetic acid, sewage sludge and varied wastewaters can be converted to hydrogen in MECs (Liu et al., 2005b; Cheng and Logan 2007; Logan et al., 2008; Pant et al., 2012). MECs can also convert the byproducts of dark fermentation such as acetate into hydrogen with high efficiencies (Liu et al., 2005b; Logan et al., 2008). This produced hydrogen can be used to generate methane in consequent reactions in special conditions. In addition, MECs require relatively low energy input (0.2-0.8 V) compared to typical water electrolysis (>2.1 V) (Zhang and Angelidaki 2014). Figure 2 demonstrates a schematic view of MEC structure.

A microbial electrolysis cell (MEC) is a bioelectrochemical system (BES) including an anode and a cathode, separated by an ion exchange membrane. At the anode, electro-active microorganisms oxidize organic or inorganic waste substrates by the electrode as terminal electron acceptor (Rabaey and Rozendal 2010). Electrons are transferred from anode to cathode and used to reduce oxidized species into valuable chemicals in the presence of a suitable catalyst. Generally, the generated potential from oxidation of substrates at anode needs to be amplified by an external power supply in order to overcome the thermodynamic barrier and/or drive the cathodic reaction at high rates. Until now, the most studied cathodic reaction in MECs is hydrogen production from protons reduction.

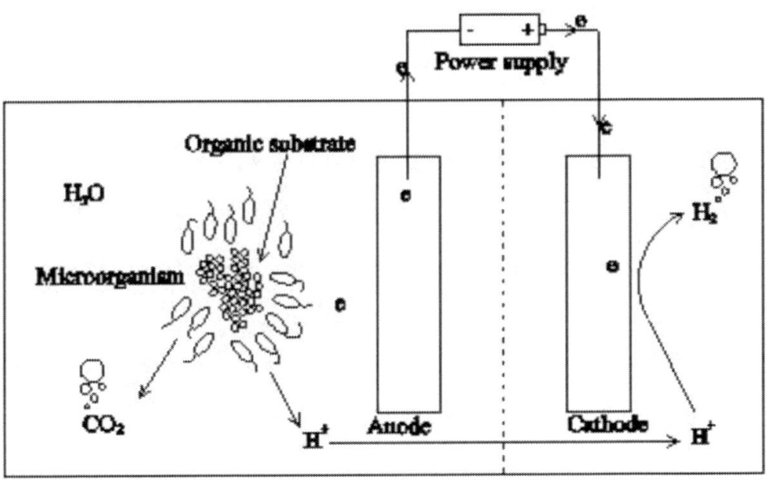

Figure 2. A schematic view of MEC structure.

This reaction usually requires the addition of an external voltage higher than 0.2 V and the presence of special metal catalysts such as platinum(Logan et al., 2008; Lu et al., 2010). Regardless of their excellent electrocatalytic activity towards hydrogen production, similar to MFCs, the considerable high price of these catalysts limits drastically their utilization in practical application. Alternative low-cost cathodic catalysts, including carbon felt, stainless steel, tungsten carbide, and nickel alloys are currently being investigated, but these materials often show insufficient chemical stability and reactivity at neutral pH (Selembo et al., 2009; Manuel et al., 2010; Call et al., 2009; Lee et al., 2009; Villano et al., 2011). In a MEC, exoelectrogenic bacteria oxidize soluble biomass and transfer electrons to the anode. At the cathode, hydrogen gas is generated through the reduction of water molecules. As mentioned above, to drive the redox reactions at MEC electrodes (i.e., oxidation of organics and reduction of water), a small amount of electric potential is required (Liu et al., 2005; Rozendal et al., 2006; Logan et al., 2008). Methanogenesis is often considered as the greatest challenge in achieving practical applications of MEC (Tice and Kim 2014). This requisite energy is substantially small compared to the minimum threshold voltage of 1.23 V for conventional water electrolysis. In a well-controlled MEC system, even if the amount of consumed energy is very low, the energy recovery in the form of hydrogen gas is substantially up to 2-4 times more than consumed electrical energy (Cheng and Logan 2007; Hu et al., 2008). This high energy recovery enables MECs to be a promising method for sustainable wastewater treatment (Logan et al., 2008). This unique potential drove substantial improvements in MEC design and operation through numerous researches. According to the recent reports on pilot-scale experiments, MECs can be practically applicable in wastewater treatment and energy recovery in the near future (Logan 2010; Cusick et al., 2011; Gil-Carrera et al., 2013; Heidrich et al., 2013).

In MEC systems energy is added by either utilizing an external power source or setting an electrode potential employing potentiostat. Hydrogen gas is generated at the cathode at a theoretical potential of –0.41 V under standard conditions and pH value of 7. Acetate oxidation by microorganisms can be attained an anode potential down to –0.28 V. Therefore, theoretically a minimum of –0.13 V is the necessary addition to the system. This voltage is less than the required value to split water (–1.2 V) because of the positive thermodynamics of organic matter degradation compared with the energy demanding reaction for splitting water. Microorganisms can be used at the cathode to transfer electrons directly or through evolved chemicals indirectly and catalyze the inorganic chemicals production. Electrotrophic microorganisms are not true catalysts because they derive energy from this process (Heidrich et al., 2013).

Methane is commonly generated in MECs, primarily from produced hydrogen gas at the cathode by acetoclastic methanogenesis process (Parameswaran et al., 2010). It is shown that methanogenic communities could directly accept electrons from the cathode (Cheng et al., 2009). Although only a small amount of Hydrogen gas is normally lost in laboratory MECs (Ambler and Logan, 2011), methanogenesis can eliminate hydrogen gas recovery in larger systems (Rader and Logan 2010; Cusick et al., 2011) unless the generated gas at the cathode is kept separated from methanogens (Nam and Logan 2011).

1.3. A Comparison of MFCs with MECs

The most important limitation of using MFC as a stand-alone technique for electrical power generation is the low amount and high cost of the produced electricity by them.

MECs offer lots of opportunities for novel applications than other processes such as anaerobic digestion, because they can produce various products. Although the electricity required in MEC systems is much lower than conventional electrolysis processes, their energy consumption is still high. Thus, a renewable source of the electricity is required to make these technologies a sustainable and clean method for hydrogen production. In order to upgrade each of these technologies, recently many researchers tried to combine them with other electrochemical processes like reverse electro dialysis (RED) systems.

1.4. Microbial Reverse Electrodialysis Cell (MRC)

The integration of MFC and RED stack into a single process, called microbial reverse-electrodialysis cell (MRC), is a new approach for energy generation. Reverse electrodialysis (RED) uses salinity-gradient power for direct electricity production, obtained from saline and hypersaline water sources, using anion and cation exchange membranes located between two electrodes. Numerous membrane pairs are required to effectively convert the salinity gradient to electricity, resulting in high costs for stand-alone RED systems (Wick 1978; Ramon et al., 2011). In this process charged counter ions to the membranes are driven from seawater to river water by the salinity gradient. This driving force quantified as the junction potential under open circuit condition and can create theoretically ~0.1 to 0.2 V per pair of anion and cation exchange membranes for various salinity ratios of seawater to river water. This potential can be amplified in proportion to the number of the cell pairs of seawater and river water so that voltage and power can be strengthened by adding more ion exchange membranes in the RED stack. For instance, a 50-cell paired RED system generates about 3 V at corresponding maximum power of 93 W/m^2 of a projected cross-sectional area of generated current (0.93 W/m^2 as normalized by total membrane area) if energy losses due to electrode overpotentials and pumping were ignored (Veerman et al., 2009). While RED is a promising technology to provide electrical power from salinity driven energy gradient, estimated to be 1.7 TW globally, practical applications have not yet been successful due to several restrictions which limit the overall efficiency of a RED system, such as:

- ohmic resistances in river water cells,
- electrode over potentials, and
- bio-fouling.

The MRC process can overcome restrictions of the individual processes because each of the combined technologies covers the other ones limitations. For example, the low voltage generated by an MFC can be amplified by salinity driven potential with the RED stack. Exoelectrogenic bacteria at the anode and oxygen reduction at the cathode drive spontaneous reactions at the electrodes in the MRC, providing additional voltage to the RED stack, compared with energy losses at the electrodes in a RED system. The integration of these two alternative energy production methods can improve power generation by removing lots of the limitations of the individual technologies (Kim and Logan 2011). However, RED and MRC applications are currently bounded to estuarine or coastal areas because of the need for both fresh and saltwater solutions.

Low-grade waste heat can be employed to generate salinity gradients (high and low concentrations of salt solutions) from thermolytic chemicals such as ammonium bicarbonate. MRCs can produce 3 W/m^2 via domestic wastewater and 5.6 W/m^2 via acetate with 30% energy efficiency using ammonium bicarbonate (not including distillation energy). Thermolytic solutions enable the use of MRCs at any location where there is wastewater or waste biomass and waste heat. The use of thermolytic solutions also minimizes biofouling of the membranes (Logan and Rabaey 2012).

1.5. Microbial Reverse Electrodialysis Electrolysis Cell (MREC)

As mentioned above, the salinity gradient power could be utilized as a source of voltage needed to enable hydrogen gas production (Kim and Logan 2011) and RED is a method for converting salinity gradient between seawater and river water into electrical power. The RED stack consists of alternating anion exchange membranes (AEMs) and cation exchange membranes (CEMs) series that dictate the direction of the flow of positive or negative ions from the high salinity solution and create a way of converting an electrochemical potential into electrical current. In a RED system, seawater and river water are pumped between the membranes in a stack that may contain 20 or more membrane pairs (about 0.1 to 0.2 V per membrane pair) to produce adequate potential to split water (Veerman et al., 2010). However, by incorporating a RED stack of only 5 membrane pairs between the electrodes in an MEC, it is attainable to both avoid the need to split water and also to eliminate the need for an external power source in order to produce hydrogen gas. This combined MEC and RED process, called a microbial reverse electrodialysis electrolysis cell (MREC). The use of the RED stack in an MEC produced H_2 gas at high energy efficiency of up to 65%, without the requirement of electrical energy (Kim and Logan 2011).

Although required electricity in MEC or RED and MRC is much lower than that of a conventional electrolysis process, the energy consumption is still high and limits their application. Thus, a renewable source of the electricity is required to make these technologies a sustainable and clean method for hydrogen production (Kim and Logan 2011). Combining a MEC and a small RED stack, represents an unsurpassed sustainable method for hydrogen production known as MREC.

In MREC, the energy for hydrogen production is derived from microbial oxidation of organic matter at the anode and the salinity gradient between seawater and river water, and thus external power resources are not required (Kim and Logan 2011). Although this technology holds great potential to act as a sustainable method to treat wastewater and simultaneously produce pure hydrogen gas without consumption of electrical energy, it is still facing several challenges such as membrane fouling with seawater. The applicability of this system in large-scale operation also needs to be more evaluated. The accessibility of both fresh and salt waters is also important for the successful implementation of this technology. Thus, the potential application of this technology is in coastal regions.

MREC systems have several limitations which confine their wide applications. One limitation of the MREC for hydrogen gas production is the required sources of organic matter and seawater close together which makes the process applicable only in coastal regions. MREC can be used in order to produce hydrogen gas in large coastal cities with wastewater treatment plants which discharge their effluent into the sea or ocean. In these cases the wastewater can first work as a source of organic matter, and then as the low salinity solution in the RED stack. Another potential restriction of these processes is the use of seawater which can result in biofouling of the membranes. To overcome area limitation and avoid fouling problems, researchers have found a novel approach via using recycled sources of clean salt solutions such as ammonium bicarbonate (NH_4HCO_3) instead of seawater (Nam et al., 2012). According to the results, MREC is able to produce enough voltage for hydrogen production using ammonium bicarbonate salts that can be easily regenerated employing low temperature waste heat (Nam et al., 2012).

Other than ammonium bicarbonate salts, magnesium sulfate, sodium sulfate, sodium chloride, potassium sulfate, potassium nitrate, potassium chloride have all been used in forward osmosis (FO) systems using osmosis pressure gradient for desalination (Achilli et al., 2010; Kim et al., 2012; McCutcheon et al., 2005). Ammonium bicarbonate is exclusive among these salts because it can easily be regenerated at temperatures below the boiling point of water (McCutcheon et al., 2006). This solution could also be used effectively as the high salt solution in a RED stack to produce electricity or in a MREC for hydrogen gas production (Nam et al., 2012).

The application area of MREC is really expanded by using ammonium bicarbonate salts. This system also proposes a new method to create energy from waste heat at low temperatures. However, the energy recovery of this technology has been reported at 10% on average and still needs to be improved to make MREC a more efficient, mature and competitive technology (Zhang and Angelidaki 2014).

2. Methane Generation in Microbial Electrochemical Technology (MET)

Methane is commonly observed in the MECs during hydrogen production due to the development of methanogens. CH_4 production from MECs is varied with inoculum, substrate and reactor configuration (Chae et al., 2010). Bruce Logan et al. (2002) have found that a microorganism, most likely Methanobacterium palustre, can form methane via electrons directly from current in addition to using hydrogen gas. This fundamental discovery is capable to be an efficient method to generate methane, the main component of natural gas, from electricity. It is also a methanogen, i.e., it can form methane from CO_2 and H_2 as an electron donor.

Methanobacterium palustre, regardless of its name, is not a bacterium and is an archaeon like other methanogens. As mentioned in chapter 1, methanogenic archaea have a unique biochemistry, utilizing cofactors and enzymes which are not commonly present in other types of microorganisms. To harvest hydrogen in MECs, methanogenesis must be inhibited. In these systems emerging of methanogens is undesired because it reduces the hydrogen production. The goal of membraneless MECs is to produce hydrogen gas; Call and Logan (2008) reported the production of methane as an undesired side product. Methane production carried on in a membraneless MEC (28 mL volume) despite exposure to the air for 30 to 45 minutes between batch-feeding cycles (Call and Logan 2008). Several approaches have been employed to inhibit the growth of methanogens in MECs (Call and Logan 2008; Clauwaert and Verstraete 2009; Wang et al., 2009).

Some suggested strategies to inhibit methanogenesis in membraneless MECs are as follows (Call and Logan 2008; Rozendal et al., 2008):

- operating under low pH conditions,
- wash-out of methanogens by a short hydraulic retention time, and
- low carbonate levels

However, most of the methods are ineffective or energy intensive.

In the other hand methane contains also high level of potential chemical energy and is an important product of anaerobic digestion (Weemaes and Verstraete 1998; Appels et al., 2008). Although the energy content per unit weight of methane is lower than that of hydrogen, methane production in MECs is believed to be more sturdy than hydrogen production (Clauwaert and

Verstraete 2009). Instead of inhibiting the methanogens, the MEC approach for methane production is particularly interesting, since it has several advantages compared to traditional anaerobic digestion processes, such as:

I. Organic matter oxidation and methane production are two separated processes in MECs: allow high methane content in biogas;
II. Methanogens accept electrons directly from cathode: makes the methanogenic consortia more tolerant against inhibitors or toxic compounds such as ammonia (Clauwaert et al., 2008) which may present in the waste material;
III. Less thermal energy (if any) is needed to control the temperature of the cathode because the process occurs at ambient temperature: since the wastewater does not need to be warmed up thereby saving energy;
IV. Diluted streams with low organic matter content can be used: including municipal wastewaters where anaerobic digestion cannot function (Villano et al., 2011).

It seems that MECs will not replace conventional anaerobic digestion processes as they are more favorable for low-strength waste streams. Therefore, MECs can be a complement to the anaerobic digestion reactor, for example, as a downstream process to polish the anaerobic digestion effluent(Zhang and Angelidaki 2014). Anaerobic digestion reactor and methane-producing MEC could operate in series due to the possibility to remove the residual organics contained in the effluent of a conventional anaerobic digester with such a bioelectrochemical system which is effective even at low substrate concentrations and can be operated at ambient temperature (Clauwaert and Verstraete 2009).

The direct transfer of electrons to methanogens suggests that it is possible to save solar or wind energies by generating methane from sources and storing it as fuel for later use. Producing methane electrochemically from CO_2 is difficult because of the noticeable overpotential, but methanogens can do it very easily. Compressing and storing hydrogen gas is too difficult comparing to natural gas and methane. There are pipelines of natural gas all over the world, and it is used in conventional engines (Booth 2009).

Most electrochemical strategies for direct methane production from CO_2 require precious and expensive metals or copper to reduce electrical overpotentials and the reactions are not sufficiently specific, resulting in production of undesired higher molecular weight carbon products that can foul the electrodes, (Spinner et al., 2012) but combined biological and electrochemical methods for methane production show great promise since they are specific for production of a single product without requiring for precious metals,(Logan and Rabaey 2012). Methanogenic microorganisms produce methane via only a limited number of organic molecules like acetate, formate, and methanol. However, some methanogens reduce CO_2 using hydrogen gas, enabling methane production using only these two inorganic chemicals. Biological methane production via direct electron transfer by microorganisms from a carbon electrode can provide a direct route of methane formation from electrical current, named electromethanogenesis, without the requirement to evolve hydrogen gas for hydrogenotrophic methanogenesis (Cheng et al., 2009). The energy demands for methane production by direct electron transfer to methanogens is theoretically less than that required via hydrogen gas.

In order to minimize operating costs for methane evolution, using more positive cathode potentials can reduce energy requirements. Avoiding the need for precious metals is also significant to make the overall bioelectrochemical process more environmentally sustainable.

Microbial biocathodes in which microorganisms are the electrocatalytic agents of the desired cathodic reaction have come into the front because of their main advantages of including self-regeneration of the catalyst, low cost, and greatest activity at neutral pH (Rozendal et al., 2007; Harnisch et al., 2010; Jeremiasse et al., 2010; Rosenbaum et al., 2011). Furthermore, due to their high adaptability and specificity, they can potentially be utilized to catalyze a wide spectrum of practical reactions. Microbial biocathodes can be employed to reduce CO_2 to methane gas via electromethanogenesis process (Cheng et al., 2009) or even multi-carbon organic compounds via microbial electrosynthesis. Biocathode overpotentials are reducable by choosing materials that are optimized to improve direct electron transfer to methanogens or by using materials that efficiently catalyze hydrogen generation at low potentials without precious metals. The application of biocathode has greatly reduced the costs of electrode catalyst in MECs.

Useful materials that have the capability to reduce overpotentials for direct electron transfer could be those that are known to go through microbial influenced corrosion by methanogens. For instance, CH_4 can be directly produced by microbial corrosion of zero-valent iron (Dinh et al., 2004). Although the exact mechanism(s) for this type of corrosion is not clear, the process may require formation of an iron sulfide crust(Enning et al., 2012). When steel corrodes, it becomes covered with an insoluble mineral layer consisting mainly of iron oxide and/or hydroxide ($Fe_xO_x[OH]_x$), and under sulfate-reducing conditions, this layer may also contain iron sulfides (Fe_xS_x). These materials accelerate the corrosion rate because they are electrically conductive, and make easy releasing of electrons compared to pure iron (Venzlaff et al., 2013). Other materials that can be used to stimulate direct electron transfer to methanogens are semi-conductive iron mineral compounds like magnetite, as they have been shown to enhance interspecies electron transfer in methanogenic multi-cultures (Kato et al., 2012).

So far, a very small number of studies have investigated the application of MEC technology for methane production (Clauwaert et al., 2009; Cheng et al., 2009; Villano et al., 2010; Villano et al., 2011; Cheng et al., 2010; Rader and Logan 2010). Therefore, fundamental information to scale up studies is not enough to date.

Conclusion

Methanogenesis and bioenergy production in bioelectrochemical systems has been investigated. MFC is a bio-electrochemical system that can be employed to produce electrical power and simultaneously treat or digest simple organic molecules, such as carbohydrates and proteins, over and above complex mixtures of organic matter of human, animal and food-processing wastewaters. The formed biofilm on the anode electrode is suitable environment for methanogens in order to produce methane using organic material. The amount of produced methane within the biofilm is minor because of restrictions of the anaerobic conditions.

According to the high cost of electricity generated by MFCs, this technology is not able to compete with other energy sources (e.g., biogas) from economic and environmental aspects in this stage. Thus, researches have been recently initiated to broad the scope of MFCs for more value-added applications by MECs like producing hydrogen gas and simultaneously various biomasses such as cellulose, glucose, glycerol, acetic acid. The direct transfer of electrons to methanogens in MECs suggests that it is possible to save solar or wind energy by generating methane from sources and storing it as fuel for later use. Producing methane electrochemically from CO_2 is difficult because of its noticeable overpotential, but methanogens are able to do it very easy.

In order to optimize the best use of each of these technologies, recently they have been combined with other electrochemical processes like the reverse electro dialysis (RED) system. The integration of a MFC and RED stack into a single process creates a new technology for energy production, called a microbial reverse-electrodialysis cell (MRC). The MRC process can overcome restrictions of the individual processes because each of them covers the other ones limitations. The combined MEC and RED reactor is called a microbial reverse electrodialysis electrolysis cell (MREC). By inserting the RED stack in a MEC hydrogen gas can be produced at high energy efficiency up to 65%, without the requirement of electrical energy that can be used for methane formation.

REFERENCES

Achilli, A. C., Tzahi, Y. & Childress, A. E. (2010).Selection of inorganic-based draw solutions for forward osmosis applications. *Journal of Membrane Science, 364*(1–2), 233-241.

Aelterman, P., Rabaey, K., Pham, H. T., Boon, N. & Verstraete, W. (2006). Continuous Electricity Generation at High Voltages and Currents Using Stacked Microbial Fuel Cells. *Environmental Science & Technology, 40*(10), 3388-3394.

Ahn, Y., Logan, B. E. (2010). Effectiveness of domestic wastewater treatment using microbial fuel cells at ambient and mesophilic temperatures. *Bioresource Technology, 101*(2), 469-475.

Ambler, J. R., Logan, B. E. (2011). Evaluation of stainless steel cathodes and a bicarbonate buffer for hydrogen production in microbial electrolysis cells using a new method for measuring gas production. *International Journal of Hydrogen Energy, 36*(1), 160-166.

Appels, L., Baeyens, J., Degrève, J. & Dewil R. (2008). Principles and potential of the anaerobic digestion of waste-activated sludge.*Progress in Energy and Combustion Science, 34*(6), 755-781.

Booth, B. (2009). Electromethanogenesis: the direct bioconversion of current to methane. *Environmental Science & Technology, 43*(13), 4619-4619.

Call, D. F., Merrill, M. D. & Logan, B. E. (2009). High Surface Area Stainless Steel Brushes as Cathodes in Microbial Electrolysis Cells. *Environmental Science & Technology, 43*(6), 2179-2183.

Call, D., Logan, B. E. (2008).Hydrogen Production in a Single Chamber Microbial Electrolysis Cell Lacking a Membrane.*Environmental Science & Technology, 42*(9), 3401-3406.

Call, D. F., Merrill, M. D. & Logan, B. E. (2009). High Surface Area Stainless Steel Brushes as Cathodes in Microbial Electrolysis Cells. *Environmental Science & Technology, 43*(6), 2179-2183.

Chae, K. J., Choi, M. J., Kim, K. Y., Ajayi, F. F., Chang, I. S. & Kim, I. S. (2010). Selective inhibition of methanogens for the improvement of biohydrogen production in microbial electrolysis cells. *International Journal of Hydrogen Energy, 35*(24), 13379-13386.

Chaudhuri, S. K., Lovley, D. R. (2003). Electricity generation by direct oxidation of glucose in mediatorless microbial fuel cells. *Nat. Biotech., 21*(10), 1229-1232.

Cheng, K. Y., Ho, G. & Cord-Ruwisch, R. (2010). Novel Methanogenic Rotatable Bioelectrochemical System Operated with Polarity Inversion. *Environmental Science & Technology, 45*(2), 796-802.

Cheng, S., Liu, H. & Logan, B. E. (2006a). Increased performance of single-chamber microbial fuel cells using an improved cathode structure. *Electrochemistry Communications, 8*(3), 489-494.

Cheng, S., Liu, H. & Logan, B. E. (2006b). Increased Power Generation in a Continuous Flow MFC with Advective Flow through the Porous Anode and Reduced Electrode Spacing. *Environmental Science & Technology, 40*(7), 2426-2432.

Cheng, S., Logan, B. E. (2007a). Ammonia treatment of carbon cloth anodes to enhance power generation of microbial fuel cells. *Electrochemistry Communications, 9*(3), 492-496.

Cheng, S., Logan, B. E. (2007b). Sustainable and efficient biohydrogen production via electrohydrogenesis. *Proceedings of the National Academy of Sciences, 104*(47), 18871-18873.

Cheng, S., Xing, D., Call, D. F. & Logan, B.E. (2009). Direct Biological Conversion of Electrical Current into Methane by Electromethanogenesis. *Environmental Science & Technology, 43*(10), 3953-3958.

Clauwaert, P., Verstraete, W. (2009). Methanogenesis in membraneless microbial electrolysis cells. *Applied Microbiology and Biotechnology, 82*(5), 829-836.

Clauwaert, P., Aelterman, P., Pham, T. H., Schamphelaire, L., Carballa, M., Rabaey, K. & Verstraete, W. (2008). Minimizing losses in bio-electrochemical systems: the road to applications. *Applied Microbiology and Biotechnology, 79*(6), 901-913.

Cusick, R. D., Kiely, P. D. & Logan, B. E. (2010). A monetary comparison of energy recovered from microbial fuel cells and microbial electrolysis cells fed winery or domestic wastewaters. *International Journal of Hydrogen Energy, 35*(17), 8855-8861.

Cusick, R. D., Bryan, B., Parker, D. S., Merrill, M. D., Mehanna, M., Kiely, P. D. & Logan BE. (2011). Performance of a pilot-scale continuous flow microbial electrolysis cell fed winery wastewater. *Applied Microbiology and Biotechnology, 89*(6), 2053-2063.

Dinh, H. T., Kuever, J., Muszmann, M., Hassel, A. W., Stratmann, M. & Widdel, F.(2004). Iron corrosion by novel anaerobic microorganisms. *Nature, 427*(6977), 829-832.

Enning, D., Venzlaff, H., Garrelfs, J., Dinh, H. T., Meyer, V., Mayrhofer, K. & Widdel, F. (2012). Marine sulfate-reducing bacteria cause serious corrosion of iron under electroconductive biogenic mineral crust. *Environmental Microbiology, 14*(7), 1772-1787.

Gil-Carrera, L., Escapa, A., Mehta, P., Santoyo, G., Guiot, S. R., Morán, A. & Tartakovsky, B. (2013). Microbial electrolysis cell scale-up for combined wastewater treatment and hydrogen production. *Bioresource Technology, 130*(0), 584-591.

Habermann, W., Pommer, E. H. (1991). Biological fuel cells with sulphide storage capacity. *Applied Microbiology and Biotechnology, 35*(1), 128-133.

Hamelers, H. V. M., Heijne, A., Sleutels, T. H. J. A., Jeremiasse, A. W., Strik, D. P. B. T. B. & Buisman, C. J. N. (2010). New applications and performance of bioelectrochemical systems. *Applied Microbiology and Biotechnology, 85*(6), 1673-1685.

He, Z., Mansfeld, F. (2009). Exploring the use of electrochemical impedance spectroscopy (EIS) in microbial fuel cell studies. *Energy & Environmental Science, 2*(2), 215-219.

Heidrich, E. S., Dolfing, J., Scott, K., Edwards, S. R., Jones, C. & Curtis, T. P. (2013). Production of hydrogen from domestic wastewater in a pilot-scale microbial electrolysis cell. *Applied Microbiology and Biotechnology, 97*(15), 6979-6989.

Harnisch, F., Schroder, U. (2010). From MFC to MXC: chemical and biological cathodes and their potential for microbial bioelectrochemical systems. *Chemical Society Reviews, 39*(11), 4433-4448.

Hu, H., Fan, Y. & Liu, H. (2008). Hydrogen production using single-chamber membrane-free microbial electrolysis cells. *Water Research, 42*(15), 4172-4178.

Jacobson, K. S., Drew, D. M. & He, Z. (2011). Efficient salt removal in a continuously operated upflow microbial desalination cell with an air cathode. *Bioresource Technology, 102*(1), 376-380.

Jeremiasse, A. W., Hamelers, H. V. M. & Buisman, C. J. N. (2010).Microbial electrolysis cell with a microbial biocathode. *Bioelectrochemistry, 78*(1), 39-43.

Kato, S., Hashimoto, K. & Watanabe, K. (2012). Methanogenesis facilitated by electric syntrophy via (semi)conductive iron-oxide minerals. *Environmental Microbiology, 14*(7), 1646-1654.

Kiely, P. D., Cusick, R., Call, D. F., Selembo, P. A., Regan, J. M. & Logan, B. E. (2011). Anode microbial communities produced by changing from microbial fuel cell to microbial electrolysis cell operation using two different wastewaters. *Bioresource Technology, 102*(1), 388-394.

Kim, B. H., Park, H. S., Kim, H. J., Kim, G. T., Chang, I. S., Lee, J. & Phung, N. T. (2004). Enrichment of microbial community generating electricity using a fuel-cell-type electrochemical cell. *Applied Microbiology and Biotechnology, 63*(6), 672-681.

Kim, J. R., Min, B. & Logan, B. E. (2005). Evaluation of procedures to acclimate a microbial fuel cell for electricity production. *Applied Microbiology and Biotechnology, 68*(1), 23-30. doi: 10.1007/s00253-004-1845-6.

Kim, T., Kim,Y., Yun, C., Jang, H., Kim, W. & Park, S.(2012). Systematic approach for draw solute selection and optimal system design for forward osmosis desalination. *Desalination, 284*(0), 253-260.

Lee, H. S., Torres, C. I., Parameswaran, P. & Rittmann, B. E. (2009). Fate of H_2 in an Upflow Single-Chamber Microbial Electrolysis Cell Using a Metal-Catalyst-Free Cathode. *Environmental Science & Technology, 43*(20), 7971-7976.

Liu, H., Grot, S. & Logan, B. E. (2005). Electrochemically Assisted Microbial Production of Hydrogen from Acetate. *Environmental Science & Technology, 39*(11), 4317-4320.

Liu, H., Logan, B. E. (2004). Electricity Generation Using an Air-Cathode Single Chamber Microbial Fuel Cell in the Presence and Absence of a Proton Exchange Membrane. *Environmental Science & Technology, 38*(14), 4040-4046.

Liu, H., Ramnarayanan, R. & Logan, B. E. (2004). Production of Electricity during Wastewater Treatment Using a Single Chamber Microbial Fuel Cell. *Environmental Science & Technology, 38*(7), 2281-2285.

Liu, H., Grot, S. & Logan, B. E. (2005a). Production of electricity from acetate or butyrate using a single-chamber microbial fuel cell. *Environmental Science & Technology 39(2),* 658-662.

Liu, H., Grot, S. & Logan, B. E. (2005b). Electrochemically Assisted Microbial Production of Hydrogen from Acetate. *Environmental Science & Technology, 39*(11), 4317-4320.

Logan, B. E., Call, D., Cheng, S., Hamelers, H. V. M., Sleutels, T. H. J. A., Jeremiasse, A. W. & Rozendal, R. (2008). Microbial Electrolysis Cells for High Yield Hydrogen Gas Production from Organic Matter. *Environmental Science & Technology, 42*(23), 8630-8640.

Logan, B. E., Hamelers, B., Rozendal, R., Schroeder, U., Keller, J., Freguia, S. & Rabaey, K. (2006). Microbial Fuel Cells: Methodology and Technology†. *Environmental Science & Technology, 40*(17), 5181-5192.

Logan, B. E., Rabaey, K. (2012). Conversion of Wastes into Bioelectricity and Chemicals by Using Microbial Electrochemical Technologies. *Science, 337*(6095), 686-690.

Logan, B. E. (2010). Scaling up microbial fuel cells and other bioelectrochemical systems. *Applied Microbiology and Biotechnology, 85*(6), 1665-1671.

Logan, B. E. (2005). Simultaneous wastewater treatment and biological electricity generation. *Water Science & Technology 52(1-2),* 31-37.

Lovley, D. R. (2006). Bug juice: harvesting electricity with microorganisms. *Nat. Rev. Micro., 4*(7), 497-508.

Lovley, D. R., Phillips E. J. P. (1988). Novel Mode of Microbial Energy Metabolism: Organic Carbon Oxidation Coupled to Dissimilatory Reduction of Iron or Manganese. *Applied and Environmental Microbiology, 54*(6), 1472-1480.

Lu, L., Xing, D., Xie, T., Ren, N. & Logan, B. E. (2010). Hydrogen production from proteins via electrohydrogenesis in microbial electrolysis cells. *Biosensors and Bioelectronics*, 25(12), 2690-2695.

McCutcheon, J. R., McGinnis, R. L. & Elimelech, M. (2005). A novel ammonia—carbon dioxide forward (direct) osmosis desalination process.*Desalination, 174*(1), 1-11.

McCutcheon, J. R., McGinnis, R. L. & Elimelech, M. (2006). Desalination by ammonia–carbon dioxide forward osmosis: Influence of draw and feed solution concentrations on process performance. *Journal of Membrane Science, 278*(1–2), 114-123.

Manuel, M. F., Neburchilov, V., Wang, H., Guiot, S. R. & Tartakovsky, B. (2010). Hydrogen production in a microbial electrolysis cell with nickel-based gas diffusion cathodes. *Journal of Power Sources, 195*(17), 5514-5519.

Nam, J. Y., Cusick, R. D., Kim, Y. & Logan, B. E. (2012).Hydrogen Generation in Microbial Reverse-Electrodialysis Electrolysis Cells Using a Heat-Regenerated Salt Solution.*Environmental Science & Technology, 46*(9), 5240-5246.

Nam, J. Y., Logan, B. E. (2011).Enhanced hydrogen generation using a saline catholyte in a two chamber microbial electrolysis cell.*International Journal of Hydrogen Energy, 36*(23), 15105-15110.

Pant, D., Singh, A., Van Bogaert, G., Olsen, S. I., Nigam, P. S., Diels, L. & Vanbroekhoven K. (2012). Bioelectrochemical systems (BES) for sustainable energy production and product recovery from organic wastes and industrial wastewaters. *Rsc Advances 2(4),* 1248-1263.

Parameswaran, P., Zhang, H., Torres, C. I., Rittmann, B. E. & Krajmalnik-Brown Rosa. (2010). Microbial community structure in a biofilm anode fed with a fermentable substrate: The significance of hydrogen scavengers. *Biotechnology and Bioengineering, 105*(1), 69-78.

Rabaey, K., Boon, N., Siciliano, S. D., Verhaege, M. & Verstraete, W. (2004). Biofuel Cells Select for Microbial Consortia That Self-Mediate Electron Transfer. *Applied and Environmental Microbiology, 70*(9), 5373-5382.

Rabaey, K., Lissens, G., Siciliano, S.D. & Verstraete, W. (2003). A microbial fuel cell capable of converting glucose to electricity at high rate and efficiency. *Biotechnology Letters, 25*(18), 1531-1535.

Rabaey, K., Rozendal, R. A. (2010). Microbial electrosynthesis — revisiting the electrical route for microbial production. *Nat. Rev. Micro., 8*(10), 706-716.

Rabaey, K., Verstraete, W. (2005). Microbial fuel cells: novel biotechnology for energy generation. *Trends in Biotechnology, 23*(6), 291-298.

Rader, G. K., Logan, B. E. (2010).Multi-electrode continuous flow microbial electrolysis cell for biogas production from acetate.*International Journal of Hydrogen Energy, 35*(17), 8848-8854.

Ramon, G.Z., Feinberg, B.J. & Hoek, E.M.V. (2011). Membrane-based production of salinity-gradient power. *Energy & Environmental Science, 4*(11), 4423-4434.

Rosenbaum, M., Aulenta, F., Villano, M. & Angenent, L. T. (2011). Cathodes as electron donors for microbial metabolism: Which extracellular electron transfer mechanisms are involved? *Bioresource Technology, 102*(1), 324-333.

Rozendal, R. A., Hamelers, H. V. M., Euverink, G. J. W., Metz, S. J. & Buisman, C. J. N. (2006). Principle and perspectives of hydrogen production through biocatalyzed electrolysis. *International Journal of Hydrogen Energy* 31(12), 1632-1640.

Rozendal, R. A., Hamelers, H. V. M., Rabaey, K., Keller, J. & Buisman C. J. N. (2008). Towards practical implementation of bioelectrochemical wastewater treatment. *Trends in Biotechnology, 26*(8), 450-459.

Rozendal, R. A., Jeremiasse, A. W., Hamelers, H. V. M. & Buisman, C. J. N. (2007). Hydrogen Production with a Microbial Biocathode.*Environmental Science & Technology, 42*(2), 629-634.

Selembo, P. A., Merrill, M. D. & Logan, B. E. (2009). The use of stainless steel and nickel alloys as low-cost cathodes in microbial electrolysis cells. *Journal of Power Sources, 190*(2), 271-278.

Spinner, N. S., Vega, J. A. & Mustain, W. E. (2012).Recent progress in the electrochemical conversion and utilization of CO2.*Catalysis Science & Technology, 2*(1), 19-28.

Tice, R. C., Kim, Y. (2014). Methanogenesis control by electrolytic oxygen production in microbial electrolysis cells. *International Journal of Hydrogen Energy, 39*(7), 3079-3086.

Veerman, J., Saakes, M., Metz, S. J., & Harmsen, G. J. (2009). Reverse electrodialysis: Performance of a stack with 50 cells on the mixing of sea and river water. *Journal of Membrane Science, 327*(1–2), 136-144.

Veerman, J., Saakes, M., Metz, S. J. & Harmsen, G.J. (2010). Reverse electrodialysis: evaluation of suitable electrode systems. *Journal of Applied Electrochemistry, 40*(8), 1461-1474.

Venzlaff, H., Enning, D., Srinivasan, J., Mayrhofer, K. J. J., Hassel, A. W., Widdel, F. & Stratmann M. (2013). Accelerated cathodic reaction in microbial corrosion of iron due to direct electron uptake by sulfate-reducing bacteria. *Corrosion Science, 66*(0), 88-96.

Villano, M., Aulenta, F., Ciucci, C., Ferri, T., Giuliano, A. & Majone, M. (2010). Bioelectrochemical reduction of CO_2 to CH_4 via direct and indirect extracellular electron transfer by a hydrogenophilicmethanogenic culture. *Bioresource Technology, 101*(9), 3085-3090.

Villano, M., Monaco, G., Aulenta, F. & Majone, M. (2011). Electrochemically assisted methane production in a biofilm reactor. *Journal of Power Sources, 196*(22), 9467-9472.

Wagner, R. C., Regan, J. M., Oh, S. E., Zuo, Y. & Logan, B. E. (2009). Hydrogen and methane production from swine wastewater using microbial electrolysis cells. *Water Research, 43*(5), 1480-1488.

Wang, A., Liu, W., Cheng, S., Xing, D., Zhou, J. & Logan, B. E. (2009). Source of methane and methods to control its formation in single chamber microbial electrolysis cells.*International Journal of Hydrogen Energy, 34*(9), 3653-3658.

Weemaes, M. P. J., Verstraete, W. H. (1998).Evaluation of current wet sludge disintegration techniques. *Journal of Chemical Technology & Biotechnology, 73*(2), 83-92.

Wen, Q., Yang, T., Wang, S., Chen, Y., Cong, L. & Qu, Y. (2012). Dechlorination of 4-chlorophenol to phenol in bioelectrochemical systems. *Journal of Hazardous Materials* 244, 743-749.

Wick, G. L. (1978).Power from salinity gradients. *Energy, 3*(1), 95-100.

Zhang Y., Angelidaki I. (2012a). Bioelectrode-based approach for enhancing nitrate and nitrite removal and electricity generation from eutrophic lakes. *Water Research, 46*(19), 6445-6453.

Zhang, Y., Angelidaki I. (2013). A new method for in situ nitrate removal from groundwater using submerged microbial desalination–denitrification cell (SMDDC). *Water Research, 47*(5), 1827-1836.

Zhang, Y., Angelidaki, I. (2014). Microbial electrolysis cells turning to be versatile technology: Recent advances and future challenges. *Water Research, 56*(0), 11-25.

Zuo, Y., Cheng, S., Call, D. & Logan, B. E. (2007). Tubular Membrane Cathodes for Scalable Power Generation in Microbial Fuel Cells. *Environmental Science & Technology, 41*(9), 3347-3353.

INDEX

A

B

C

D

E

F

G

H

I

N

O

P

R

S

T

U

V

W